SpringerBriefs on Cyber Security Systems and Networks

The series aims to develop and disseminate an understanding of innovations, paradigms, techniques, and technologies in the contexts of cyber security systems and networks related research and studies. It publishes thorough and cohesive overviews of state-of-the-art topics in cyber security, as well as sophisticated techniques, original research presentations and in-depth case studies in cyber systems and networks. The series also provides a single point of coverage of advanced and timely emerging topics as well as a forum for core concepts that may not have reached a level of maturity to warrant a comprehensive textbook. It addresses security, privacy, availability, and dependability issues for cyber systems and networks, and welcomes emerging technologies, such as artificial intelligence, cloud computing, cyber physical systems, and big data analytics related to cyber security research. The mainly focuses on the following research topics:

Fundamentals and theories

- Cryptography for cyber security
- Theories of cyber security
- Provable security

Cyber Systems and Networks

- Cyber systems security
- Network security
- Security services
- Social networks security and privacy
- Cyber attacks and defense
- Data-driven cyber security
- Trusted computing and systems

Applications and others

- Hardware and device security
- Cyber application security
- Human and social aspects of cyber security

More information about this series at http://www.springer.com/series/15797

Dietmar P. F. Möller

Cybersecurity in Digital Transformation

Scope and Applications

 Springer

Dietmar P. F. Möller
Computer Engineering
Clausthal University of Technology
Clausthal-Zellerfeld, Germany

ISSN 2522-5561   ISSN 2522-557X   (electronic)
SpringerBriefs on Cyber Security Systems and Networks
ISBN 978-3-030-60569-8   ISBN 978-3-030-60570-4   (eBook)
https://doi.org/10.1007/978-3-030-60570-4

This Springer imprint is published by the registered company Springer Nature Switzerland AG
The registered company address is: Gewerbestrasse 11, 6330 Cham, Switzerland

Foreword

Modern market dynamics are reviving up the industry. New industry is turning more to the combine of automation and information technology, in order to improve quality, productivity, safety, speed, competitiveness, flexibilities, and reducing costs. The recent innovations on digital transformation have been at the forefront in setting a new industrial model, which is named as fourth Industrial Revolution in Germany today.

The German Industry 4.0 raised new intelligent applications, such as intelligent manufacturing, powered by different advanced technologies as intelligent and connected components from the cyber and the physical world, which have to be designed by security measures to prevent cyber threat attacks by cyber-criminals. Beside this, security against cyber-criminal threat attacks is an essential issue in the data processing through various intelligent network and communication systems.

This book shows the importance and the basic concept of cybersecurity, a systematic overview of the latest development in methods and technologies, and provides a whole framework of cybersecurity in digital transformation. The concepts of threat intelligence, intrusion detection and prevention, machine learning and deep learning, attack models and scenarios, cybersecurity ontology, and challenges for cybersecurity leadership are introduced, and discussed in detail. Author has also analyzed the threat intelligence, how to prevent threat intrusion detection, system complexity under varying constrains and scenarios in intelligent manufacturing.

In this context, the Sino-German Lighthouse Cooperation project TEDUNET with its focus on Intelligent Manufacturing, supported by the Chinese Ministry of Industry and Information Technology (MIIT), should be mentioned. The TEDUNET project is between Zhengzhou University (ZZU), China, and Technology University Clausthal (TUC), Germany, and focus on CPS, cybersecurity in intelligent manufacturing. Cybersecurity and intelligent manufacturing are two important issues in Chinese national policy. I have studied many from the draft of this book!

I think this book has shown many creative ideas of ongoing research work and fundamentals focusing on cybersecurity in digital transformation, which are derived from the practices by authors in TU Claustahl. The chapters of this book are well written and organized, showing the deep understanding thorough professional knowledge about the cybersecurity. Therefore, I would like to strongly recommend: this book should be looked as a very important book or textbook to gain knowledge in cybersecurity in digital transformation.

前 言

不断变化的市场动态正在振兴整个行业。因此，工业正在将自动化作为提高质量，生产率，安全性，速度，竞争力和降低成本的一种手段。不断发展的技术投资环境对价值链合作，内部运营以及产品和服务的客户体验在不久的将来将如何产生产生深远的影响。最新的技术创新始终在树立新范式方面走在前列，如今已被誉为通过数字化转型带来的第四次技术进步。这种范式提出了新的智能应用程序，例如智能制造，这些智能应用由不同的先进技术提供支持，例如来自网络和物理世界的智能和互连组件，必须通过安全措施来设计它们，以防止网络威胁和犯罪。除此之外，通过具有无限可能的各种基于智能信息技术的系统进行数据处理是一个至关重要的问题，其针对网络犯罪的网络威胁攻击的安全性也可以保证系统不受干扰。因此，此专著展示了正在进行的研究工作的创新思想和针对数字化转型背景下的网络安全的基础知识，这是德国克劳斯塔尔工业大学与中国郑州大学之间的研究合作的一部分。

该专著展示了如何分析数字化转型时代在各种操作条件和场景下固有的系统复杂性，以预测智能制造中系统和网络的潜在安全风险。在这种情况下，应提及由中国工业和信息技术部（MIIT）批准的中德智能制造合作示范项目 TEDUNET。 TEDUNET 项目是中国郑州大学（ZZU）和德国克劳斯塔尔工业大学（TUC）之间合作项目，其重点是智能制造中的网络物理系统及其网络安全。网络安全和智能制造也是中国国家议程中的两个重要问题。

专著的章节写得很好，显示了作者的学术严谨性和专业性。因此，该专著是获取数字化转型背景下的网络安全知识的重要读物。

刘炯天教授

郑州大学校长，中国

Sino-German Institute, Zhengzhou University   Weiyan Hou
Zhengzhou, Henan, China

Preface

The goal of this monograph is to provide a comprehensive, in-depth, and state-of-the-art summary of cybersecurity in the era of digital transformation, which has achieved great interest recently, particularly in public and private organizations. The monograph provides a systematic overview of the latest development in methods and technologies in cybersecurity in the era of digital transformation with the focus on cybersecurity, threat intelligence, intrusion detection and prevention, machine learning and deep learning, attack models and scenarios, cybersecurity ontology, and challenges for cybersecurity leadership. Thus, the monograph provides a framework within which the reader can assimilate the associated requirements. Without such a reference, the practitioner is left to ponder the plethora of terms, standards, and practices that have been developed independently and that often lack cohesion, particularly in nomenclature and emphasis. Therefore, this monograph is intended to both cover all aspects of cybersecurity in digital transformation and to provide a framework for consideration of the many issues associated with cybersecurity in digital transformation. The outline is as follows:

Chapter 1 describes the importance of cybersecurity in the era of digital transformation with regard to the impact of the fourth technological wave and their importance to circular economy.

Chapter 2 provides the relevant cybersecurity background which is among the most important board-level issues for nearly every public and private organization. Thus, the protection of data, generated in connected digital transformative environments, accessible from any location and at any time, is important in order to stay secure through cybersecurity methods.

Chapter 3 introduces threat intelligence as an evidence-based knowledge that allows preventing or mitigating cyber threat attacks to data. Threat intelligence results in methods of informed decisions about security by answering questions such as who is attacking, what is attacker's motivation as well as capability, and others. This requires a detailed analysis of well-known and documented attack incidents, which may reveal identifiable cyber threat interactions or dependency patterns.

Chapter 4 discusses threat intrusion, detection, and prevention methods to identifying malicious incidents, logging information about them, attempting to stop them and reporting the identified malicious attack to incident response teams for support. In this regard, intrusion detection and prevention strategies are becoming important knowledge to decide about the right approach to secure critical and crucial infrastructure against malicious cyber threat attack incidents.

Chapter 5 surveys machine learning and deep learning, two methodologies which have gained importance due to the impact of digital transformation and the increasing growth of data sets, to improve intrusion detection system performance. Two use case examples are introduced.

Chapter 6 provides knowledge about cyber attackers, to develop attack models for security analysis of different cyber-attack scenarios. This allows simulating distinct attack paths or alternative approaches on how to secure the targets the cyber attacker tries to attack. This methodology is introduced as a profiling approach to gain prior knowledge for potential cyber-criminal attacker scenarios, based on cyber threat attackers' motivation.

Chapter 7 discusses ontologies which are explicit conceptualizations of subject domains and therefore an important method for knowledge acquisition and knowledge sharing. Ontologies are an essential methodological approach for knowledge-intensive problem-solving that involves reasoning about objects and concepts in a particular domain or information resource with the scope enhancing cybersecurity in the respective data space.

Chapter 8 concludes this monograph by providing an overview of the challenges to gain the respective knowledge for cybersecurity leadership in the era of digital transformation.

I thank all authors who have published cybersecurity material and directly or indirectly contributed to this monograph through citation. In particular I thank the president of TU Clausthal, Germany, Prof. Dr. Joachim Schachtner, for supporting my research work on the fourth technological wave. This research is partly supported by "Study of the Sino-German Joint-Training-Model for Engineering Students of the Sino-German Lighthouse Cooperation Projects on Intelligent Manufacturing" of the Chinese Ministry of Industry and Information Technology (MIIT), and the "Strategic Consultant Research Project of the Henan Sub-branch of the Chinese Academy of Engineering 2020." I also would like to thank Mr. Ashwin Bala Vidya of QSO-Technologies India Pvt. Ltd. for his excellent assistance in proof reading. Most notably I would like to deeply thank my wife Angelika, my daughter Christina, and my grandchildren Hannah, Karl, and Teresa, for their encouragement, patience, and understanding while I was writing the monograph.

Clausthal-Zellerfeld, Germany   Dietmar P. F. Möller

Disclaimer: The information in this book provides general information about cyber-security and digital transformation as guidance; it is neither intended as legal advice nor should anybody consider it as such.

Acronyms

AA	Application awareness
AABM	Adversary attack behavior model
A_{cc}	Accuracy
ACG	Assumptions, capabilities, goals
AES	Advances Encryption Standard
AI	Artificial intelligence
AID	Anomaly intrusion detection
AIDS	Anomaly intrusion detection system
AIPSA	Anomaly-based IPS architecture
ANN	Artificial neural network
API	Application programming interface
APT	Advanced persistent threat
BC	Business continuity planning
BM	Blacklist-based method
C_A	Cyber attacker
CAPEC	Common attack pattern enumeration and classification
CC	Correlation coefficient
CDT	Cognitive-based detection techniques
CERT	Computer emergency readiness teams
CfR	Classification rate
CIA	Central Intelligence Agency
CID	Calculating the intrusion detection capability
CIF	Collective intelligence framework
CFLM	Classification learning method
CLASP	Comprehensive lightweight application security process
CLM	Clustering learning method
CM	Cognitive model
COA	Center of average method
COG	Center of gravity method
CPE	Cost per example
CR	Classification rate

CRIT	Collaborative research into threats
Cs_r	Credence
CSL	Cybersecurity leadership
CSV	Comma separated value
C_{TA}	Cyber threat attack
CTA	Cyber threat attacker
CTI	Cyber threat intelligence
CTIA	Cyber threat intelligence algorithms
C2	Command-and-control
C2IP	C to integer program
CVE	Common vulnerabilities and exposures
CWE	Common weakness enumeration
CWSM	Common weakness scoring system
CyboX	Cyber observable expression
CyR	Cyber resilience
DADS	Dynamic anomaly detection system
DAML	DARPA Agent Markup Language
DBN	Deep belief network
DBNN	Deep Bayesian neural network
DCM	Density-based clustering method
DDN	Deep directed network
DDoS	Distributed denial of service
DL	Deep learning
DLO	Domain-level ontology
DNS	Domain name servers
DO	Domain ontology
DoS	Denial of service
DPI	Deep packet inspection
DR	Disaster recovery
DSO	Domain ontology
DT	Digital transformation
DTM	Decision tree method
DtR	Detection rate
ELM	Ensemble learning methods
ES	Expert system
E2E	End-to-end
FAR	False alarm rate
FD	Fraud detection
FFANN	Feed-forward artificial neural network
FIIPS	Firewall integrated intrusion prevention system
FIPD	Forensic investigation profile document
FNPV	False negative prediction value
FO	Foundation ontology
FPPV	False positive prediction value
FNR	False negative rate

FPR	False positive rate
GDPR	General data protection regulation
GOA	U.S. Government Accountability Office
HCM	Hierarchical clustering method
HIDS	Host-based intrusion detection system
HIPS	Host-based intrusion prevention system
HTTPS	Hyper-text transfer protocol secure
ICMP	Internet control message protocol
ICSS	Industrial control system security
ICT	Information and Communication Technology
ICMP	Internet control message protocol
IDP	Intrusion detection process
IDPS	Intrusion detection and intrusion prevention system
IDPSA	Intrusion detection and prevention system architecture
IDS	Intrusion detection system
IDSA	Intrusion detection system architecture
IoC	Indicators of compromise
IoCL	Indicator of compromise lifecycle
IoE	Internet of everything
IoT	Internet of things
IIoT	Industrial internet of things
IP	Internet protocol
IPS	Intrusion prevention system
IPSA	Intrusion prevention system architecture
IT	Information Technology
KK	Known-known
KU	Known-unknown
LAN	Local area network
LDA	Linear discriminant analysis
MAEC	Malware attribute enumeration and characterization
MANTIS	Model-based analysis of threat intelligence source
MDP	Markov decision process
MID	Misuse intrusion detection
MISP	Malware information sharing platform
MITRE	Massachusetts Institute of Technology Research
ML	Machine learning
MLO	Mid-level ontology
MO	Meta-ontology
MOM	Mean of maximum method
MRS	Model-based reasoning system
MVKC	Minimum viable kill chain
NFGW	Next generation firewall
NFR	Network flight recorder
NIDS	Network-based intrusion detection system
NIPS	Network-based intrusion prevention system

NNM	Nearest neighbor method
NRL	Negative reinforcement learning
NVD	National vulnerability database
MwF	Malware filtering
OASIS	Open standards open source
OpenIoC	Open indicators of compromise
OS	Operating system
OSS	Operating system scheduler
OTA	Over-the-air
OWASP	Open web application security project
OWL	Web ontology language
P	Precision
PbCM	Probability-based clustering method
PERT	Program evaluation and review technique
PFF	Packet-filtering firewalls
PIPSA	Policy-based IPS architecture
POTS	Plain old telephone service
PRL	Positive reinforcement learning
PRLA	Policy-based reinforcement learning algorithm
PtCM	Partitioning clustering methods
P2P	Peer-to-peer
QoS	Quality of service
QL	Q learning
RBM	Restricted Boltzmann machine
RDF	Resource description framework
RDMS	Rational database management system
RIDS	Rule-based intrusion detection system
RLM	Regression learning method
R_{LM}	Risk level
R_{LAM}	Marginal risk level
R_{LAE}	Actual executed risk level
RML	Reinforcement machine learning
RMP	Reputation-based malware protection
RMSE	Root mean square error
RO	Representation ontology
S	Security
S_C	Source
SADS	Static anomaly detection system
SCADA	Supervisory control and data acquisition
SCAP	Security content automation protocol
SEP	Soltra edge platform
SID	Signature intrusion detection
SIDS	Specification-based intrusion detection system
SIDT	Specification-based intrusion detection techniques
SIPSA	Signature-based IPS architecture

SM	Statistical model
SML	Supervised machine learning
SNI	Server name indication
SOF	Standardized output formats
SOM	Self-organizing maps
SPT	Security penetration test
SQL	Standard query language
SSH	Secure shell
SSL	Secure socket layer
STIX	Structured threat information eXpression
TAXII	Transport protocol automated eXchange of indicator information
TCP	Transmission control protocol
TIDS	Threat intelligence defense system
TIM	Threat intelligence model
TIMP	Threat intelligence management platform
TNR	True negative rate
TPR	True positive rate
TPPV	True positive prediction value
TNPV	True negative prediction value
TSM	Time series model
TTE	Techniques, tactics, exploits
TTP	Tactics, techniques, procedures
UCO	Unified cybersecurity ontology
UDP	User datagram protocol
ULO	Upper level ontology
UML	Unsupervised machine learning
UO	Upper ontology
UU	Unknown-unknown
VERA	Violent extremist risk assessment
VRLA	Value-based reinforcement learning algorithm
VOIP	Voice over IP
VPN	Virtual private network
WASC	Web Application Security Consortium
W3C	World Wide Web Consortium
WM	Whitelist-based method
XML	Extensible markup language

Contents

Chapter 1
Introduction to Digital Transformation

1.1 Digital Transformation

The cyber world is an endlessly expanding space which offers huge opportunities for the digital transformation due to the existing high cyber potential and interconnectivity. In this space the raw material and hence the basis of digital transformation is data. This can be seen in the amount of data, for instance in industrial applications, that rises to gigantic amounts through the rapid growth in digital technologies such as Computing Technologies, Information and Communication Technology, Wireless Connectivity, Sensor and Actuator Nodes, the Internet, Artificial Intelligence, Cloud Computing, Machine Learning [1, 2], and many others. This drives the evolution and challenges of the digital age and hence the digital transformation. The term digital comes from the Latin word *digitus* and refers to one of the oldest forms of counting in the analog physical world. Therefore, if information is transmitted, forwarded or stored in a digital format, it is converted into numbers, at the most basic machine level as zeroes and ones (0, 1) so that computers can process, transmit, and store such information. Against this background, the word digital can also be used as an indicator for the change occurring in today's world, referred to as cyber-physical world, driven by the rapid adoption of digital technologies, where the cyber and the physical worlds are partly overlapping. Albeit, the cyber world is continuously evolving over time, all of the details of the cyber world cannot be known by everybody or at a certain moment of time. Thus, digital transformation refers to the integration of the digital technology of the cyber world into all physical domains like industry, and others, fundamentally changing operation and delivering value to customers. For instance, industrial additive manufacturing, also known as 3D printing, consists of constructs and items of the cyber and physical worlds, referred to as a cyber-physical world that are dependent on each other, within the 3D printing cyber-physical system.

D. P. F. Möller, *Cybersecurity in Digital Transformation*, SpringerBriefs on Cyber Security Systems and Networks, https://doi.org/10.1007/978-3-030-60570-4_1

Besides this, there is also a social change happening due to the existing high cyber potential and interconnection. New players in the cyber world are emerging and attracting people, for example online shopping and social media, and others. These emergent global players will bring a lot of cultural issues with a variety and even divergence of values to compete in the minds of people [3]. This also requires public and private organizations to continually challenge the status quo, and experiment, as well as get comfortable with the future. However, to create new sustainable and competitive strategic plans for public and private organizations, operational changes that make the evolution and challenges of the digital age and hence the digital transformation possible, has to be taken into account so that the added value moves away from the strongly linear to a networked form. Thus, in addition to the internal measures, maintaining an ecosystem and building networks is of particular importance for the success of the digital transformation which is achieved by transforming agile innovation approaches to solve complex problems, enabled by digital technologies.

However, the digital transformation involves a more sophisticated variety of advanced and intelligent technologies and skills to understand, develop, and dominate them to make business, governmental, industrial, and society processes more innovative, intelligent and efficient. Therefore, the idea behind digital transformation is to use digital technology not just to replicate an existing process in a digital form, but to use digital technology to transform that process into something intelligent, where anything is connected with everything at any time and accessible, controllable and finally significantly designable in an advanced manner. Hence, advanced competences in digital systems and digital network processes as well as deep knowledge in digital technologies are essential to dominate the digital transformation. These advanced competences must be made accessible, available, and known, as essential scopes in digital transformation.

It may be nearly impossible to know how this innovation will look like at the end of this evolutionary step, called digital transformation. However, it is the process of rapid innovations, constant learning through experience, and reiteration along the way gathering expertise which make the difference in gaining the respective knowledge in digital transformation. As described in [4] "companies and organizations that figure out how to breathe big data, how to harness the power of this new resource and extract its value by leveraging the cloud, artificial intelligence, and Internet of Things, will be the next to climb out of the data lake and master the new digital land". This include awareness and knowledge in innovative technologies in the digital transformation such as additive manufacturing, augmented and virtual reality, autonomous robots, big data and analytics, cloud edge and fog computing, cyber-physical systems, cybersecurity, intelligent manufacturing, digital twin, Industrial Internet of Things, ubiquitous computing, 5G, 6G, and many others. Moreover, the availability and accessibility of other intelligent innovative technologies such as artificial intelligence, deep learning, human-machine-interface, machine learning, machine-to-machine communication, and others, which have to be taken into account to gain the respective essential knowledge. However, while digital transformation is one of the most commonly-used phrases in private and public scenarios, definitions vary. What everyone can agree upon is that digital transformation will initiate the most essential changes to business and industrial processes and society behavior. Hence, the impact of digital transformation results in changing traditional isolated processes into fully integrated and connected data flow driven processes across borders, with regard to their self-aware, decentralized and self-optimizing systems and components. This requires wireless End-to-End (E2E) technologies for digital connectivity in information-driven real-time cyber-physical systems, for instance in intelligent manufacturing environments to perform efficiently. For this purpose, a key technology required is 5G, the fifth generation high speed and low latency wireless communication technology, a new standard recently established worldwide, through which big data, generated by connected and collaborative innovative and intelligent systems and their environments, as well as the Industrial Internet of Things, assessable and operable in real-time.

It is assumed that the challenges of digital transformation towards autonomous decision support and decision-making approaches is not only a technological shift, but also an organizational change at the intersection of technology, business, governments, and society, which show that technology in itself does not equate to digital transformation. Thus, digital transformation on the one hand requires intelligent and connected components from the cyber and the physical world which are referred to as cyber-physical systems [5], which have to be designed by security measures to avoid cyber threat attacks intruded by cyber-criminals, and on the other hand processing data from various intelligent information and operational technology based systems with its endless possibilities, which also have to be secured against cyber threat attacks. Thus cyber-physical systems must be secure against cyber threat attacks to allow undisturbed system operation.

Besides this, new business models like "as-a-service" models are used, that were unimaginable years ago. However, at present it can be stated that the digital transformation and thus the required measures for the implementation of system and network cybersecurity is not as well understood to defend against all cyber threat attacks, and a number of myths obscuring the path realizing its assumed potential for value creation in digital transformation and cybersecurity. In this regard digital transformation can be understood as a change process which proceeds due to its intrinsic dynamic development at high speed. Against this background, digital transformation is a process of change that goes hand in hand with high speed innovation cycles due to the inherent dynamic development of digital technologies and, at the same time, paves the way for further innovative technologies through existing technologies which can be described as a technological domino effect. In addition to short digital technological innovation cycles, such as those seen in the development, for instance, of smartphones, the driving force behind the digital transformation is also the change in customer requirements. This can only be served by using advanced digital cyber-physical technologies. An example of this is the Spotify streaming service, which makes music accessible anytime, anywhere, and at low cost.

To adapt to the digital transformed economy the capacity for sensing challenges and opportunities as well as for fast adapting processes and models in business, governmental, industrial as well as society organizations is essential. This also requires answering the question on how the new developed model is aware about protecting data, because cybersecurity is the most critical and crucial issue in digital transformation security to avoid cyber threat attacks. However, it will take some time to implement the digital transformation in industry, business, government and society, based on artificial intelligence and machine learning driven new business models and processes, networked intelligent machines, augmented product reality, data collection and management systems, and many others. This also requires appropriately practical and technological knowledge and competences, essential to lead digital transformation at the respective level. All in all this is a very essential high level intrinsic aspect because none of the previous technology driven waves has had a truly disruptive potential like the digital transformation wave. Against this background the digital transformation with its disruptive effects will not leave one stone on the other. Indeed, largely implemented it will show disruptive innovations. In contrast, evolutionary ones not only substitute solutions, which will create new markets and business models but also change the social life of society, as the internet has changed it. Therefore, the digital transformation with its clear essential need for continuous innovations in digital technology and cybersecurity awareness can be understood as a continual change in progress. However, the digital transformation will look different for every enterprise and public and private organization, but it will be accepted in a way that digital transformation will drive the integration of digital technology into all areas of a business or public and private organization, resulting in fundamental changes of how businesses operate and organizations work, and how they deliver values to customers. Therefore, the digital transformation wave connects anything with everything, and allows accessing and controlling anything, whereby everything will be recordable and programmable. However, there exists no clear roadmap showing what should be done first, second and so on, and what the accelerators are and what the barriers [6].

1.2   Cybersecurity

An analysis of the effect of digital transformation by advanced digital technologies with respect to the resulting changes in public and private organizations requires an overview at the entire intrinsic complex level with its intelligent and interconnected systems, devices and networks used to fulfil the respective work of public and private organizations. Therefore, analyzing the impact of advanced digital technologies in public and private organizations requires extensive technological and sociological research with regard to the interaction of advanced digital technologies as well as their cybersecurity issues, which will become an intrinsic risk through cyber threat attacks. Thus, cybersecurity as a computing-based discipline deals with the presence of adversaries and hence cyber threat attacks. Within computer science, the area of cybersecurity spans many areas, including (but not limited to) data security, cryptography, software and hardware security, network and systems security, privacy, and many others. Thus, cybersecurity is fundamental in the cyber space to both, protecting secret data and information and enabling their defense, whereby the cyber space is an artificial entity formed by bits. Against this background, cyber threat attacks become reality using advanced digital technologies with their extreme interconnected capability. Therefore, cybersecurity can be defined as a body of knowledge with regard to technologies, processes, and practices designed to protect computer systems, networks, or programs, as well as data of the cyberspace from attack, damage, or unauthorized access. In this regard the elements of cybersecurity include for instance [5]:

- *Application Security*: Security measures at the application level to make applications more secure by finding, fixing, and enhancing the security of applications, to prevent data or code within the application from being compromised, corrupted, stolen or hijacked. Much of these attacks happen during the development phase, but it includes tools and methods to secure apps once they are deployed. This is becoming more important as hackers increasingly target applications with their attacks.
- *Computer Security*: Protection of computer system or network from harm, theft, and unauthorized use as well as from disruption or misdirection of services they provide. It is becoming important due to increased reliance on computer systems and the growth of smart devices. Computer security refers to four major cyber threats:

  - Fraud
  - Invasion of privacy, such as illegal accessing of protected data
  - Theft of data
  - Vandalism or destruction of data by a computer virus

- *Data Security*: Set of standards and technologies to prevent unauthorized access to computer systems, databases websites, and others to protect data from intentional or accidental destruction, data corruption, modification or disclosure. Data security is an essential topic of IT security for public and private organizations supporting use of techniques and technologies, including administrative controls, physical security, logical controls, organizational standards and other safeguarding techniques to minimize or avoid access to unauthorized or malicious users or processes.
- *Disaster Recovery and Business Continuity Planning*: Similar practices with the goal to limit the cyber threat risk and get public and private organizations running their business tasks as close as possible to normal after an unexpected incident. As cyber threats increase and the tolerance for downtime decreases, disaster recovery and business continuity planning gain importance. Therefore, there is a continuing trend to combine disaster recovery and business continuity planning.
- *Information Security*: Strategy designed to protect the confidentiality, integrity and availability (CIA Triad of information security—see Chap. 2) of computer systems data or network data packets from those with malicious intentions. Information security includes a set of strategies for managing the processes, tools and policies necessary to prevent, detect, document and counter cyber

threats to digital and non-digital information. Thus, information security is the practice preventing unauthorized access, destruction, disclosure, disruption, inspection, modifications, or use of unauthorized information.

- *Network Security*: Term that covers a multitude of technologies, devices and processes to secure networks. The term refers to a set of rules and configurations designed to protect the confidentiality, integrity, and accessibility of computer systems or networks as well as data using both software and hardware technologies. Today's network architecture is complex and faced with a threat environment that is always changing and cyber threat attackers that are always trying to find and exploit vulnerabilities. These vulnerabilities can exist in a broad number of areas, including devices, data, applications, users and locations. Thus, network security is the practice preventing and protecting against intrusion incidents into corporate public and private networks.

Nevertheless, one of the most problematic elements of cybersecurity is the fast and constantly evolving nature of cyber threat attack risks. The traditional approach has been to focus most of the resources on the most critical and crucial processes and protect against the biggest known threats, leaving some less important systems or components undefended and systems or components exposed to some less dangerous risks. Such an approach is insufficient for the current advanced digital technology based complex public and private organizations computer systems or network environments. Against this background, cybersecurity professionals assume that the traditional approaches to securing public and private computer systems or networks information can become unmanageable because the cyber threat attack environment has become impossibly complex so that manual and semi-automated cybersecurity checks and interventions cannot keep up with a constantly evolving cyber threat attack landscape. Hence, public and private organizations computer systems or networks have been identified as vulnerable to cyber threat attacks because of their omnipresent accessibility and connectivity which makes them vulnerable to remote attacks. Against this background cybersecurity teams worldwide are trying to analyze vulnerabilities in order to gain deeper knowledge about them to build up upon efficient cybersecurity strategies for defending cyber threat attacks.

Computer systems or networks perfectly integrate computation with real world processes and provide abstractions, modeling, analysis and design techniques for their overall advanced digital technology-based conceptualization. Their integrated computational and physical capabilities interacting Over-the-Air (OTA) wireless, connecting the cyber space with the physical real-world systems and processes through new modalities in the era of digital transformation fulfilling their dedicated tasks. However, the consolidation of cyber and physical components within the digital transformation enable new categories of vulnerability with regard to interception, replacement, or removal of information from the communication channels resulting in malicious attempts by cyber threat attackers to capture, disrupt, defect, or fail the computer system or network operations. The reason for this new vulnerability can be traced to the way in which the cyber and the physical components of computer systems or networks are integrated. In this vulnerable space, the cyber component provides computing capability which means processing, control software and sensory support as well as facilitates the analysis of data received from various sources and the overall operational ability of the public and private organizations' computer systems or networks.

As described in [7] the remote network access facilitates highly-productive interaction among the various physically distributed or concurrent collaborating units of computer systems or networks, and the efficient system administration is an integral part of the cyber component through being accessible. However, this accessibility provides an entrance for launching cyberattacks. These cyber threat attacks not only have tremendous impact on the cyber part but the overall cyber and physical system components. Hence, the defense against cyber threat attacks is an essential must. This will also have an unimaginable impact on human lives besides the technological impact which is a very important issue avoiding social conflict in the era of digital transformation based future society. But this is not part of this brief which specifically focuses only on digital transformation issues through usage of digital technologies and the resulting topic of cyber threat attacks, but people have to be aware about the intrinsic sociocultural problem. However, as a result of digital transformation in intelligent manufacturing in the near future, it will be necessary to think to what extent parts of intelligent machine work should be replaced by human labor again, avoiding sociocultural conflicts in the near future. As described in [8], in the history of humanity, culture served life and technology survival. Today, technology determines lives, but what culture ensures survival? Hereupon, a balanced answer must be found soon, so that technological progress can also become sociocultural progress. Experts say that there have probably been fewer changes in the last 300 years than the next 30 years will bring. This unprecedented change will have a huge impact on the entire human life. In this regard the overall technological impact of the digital transformation process of public and private organizations depends on several interactions as schematically shown in the diagram in Fig. 1.1.

1.3   Fourth Technological Wave

The fourth wave of technological advancement through digital transformation, raised new intelligent methods, powered by technologies such as

- Artificial Intelligence
- Internet of Everything
- Cyber-Physical Systems
- Cloud Computing
- Internet of Things
- 5G
- And many others

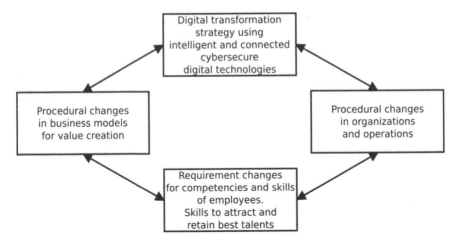

Fig. 1.1 Technological impacts of the digital transformation on business processes

These technologies will be enhanced, for instance, to make designs for manufacturability with a low carbon footprint and low resource usage by intelligent autonomous acting robots and other intelligent processes, high bandwidth networks, and others. However, this requires enablers dealing with digital technologies and digital infrastructures, such as standards, protocols, and others, developing value added supply chains based on which better business models can be set up. Against this background the fourth wave of technological advancement will also be a first wave of environmental footprint using data analytics, intelligent manufacturing, intelligent sensors, artificial intelligence and others are technologies to run more sustainable businesses. The result of what is possible will be unprecedented. As an example, for the textile industry, a company just announced new manufacturing capability that automates part of the jeans production process, allowing the company to tailor supply to meet demand, reducing textile waste and eliminating thousands of chemical substances previously needed for finishing. Also block chain applications for food supply chains are used as a part of the first wave of environmental footprint. The upside for the environment is less food waste. For business, the upside is increased supply chain efficiencies and ensured food safety as well as cutting tons of carbon dioxide from its global supply chain [9].

These new business models will be proved by industry within pilot projects, gaining experience within a proof of concept with regard to the level of complexity and machinery intelligence for data processing. In the context of machinery intelligence, different levels of intelligence can apply from information handling to problem notification to decision making. Intelligent machines and robots at the highest level make the processes independent, shaping their data based on autonomous decisions. In the context of intelligence location, one can distinguish between the intelligence of objects like machines, robots and others, and the intelligence through networks. This requires managing the respective technological investments for future fitness of the public and private to cope with high competition, demanding customers, reduced lead times and a higher degree of flexibility in industrial business, governmental and social environments. This envisioned future state of public and private organizational environments needing a structured digital transformation which is based on many technological, economical, organizational and legal challenges integrating the consequences of digital transformation to the interdependencies and challenges of governments, human workforce, industrial, science, society, and environmental issues, shown in Fig. 1.2. Nevertheless, it should be mentioned that all technological advances bear the risk of cyber threat attacks, those which require a special attention to cybersecurity.

1.4 Circular Economy

The concept of circular economy is considered as an ecological and economical model of the future. Circular economy supports public and private organizations in preserving values, conserving resources and reducing the greenhouse effect. The traditional business models in the three previous technological waves often treat resources as infinite, but awareness of today's anthropogenic effects such as climate change, pollution, and water scarcity show that this is no longer a suitable and sustainable model. Thus, legislators and consumers are therefore increasingly pushing for the sustainable use of natural resources. Against this background, the digital transformation of the fourth technological wave in conjunction with the circular economy provides essential solutions required. Circularity therefore requires both, on the one hand material and energy change which is disruptive and not only changing processes but also affecting products and business models, and on the other hand product design and material marking based digital tagging and tracking to

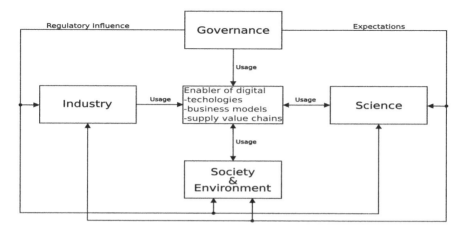

Fig. 1.2 Consequences of digital transformation

generate core data records for a sustainable product and material data space with the intention to support final disposal and recycling of valuable materials used. The resulting circular economy product and material data space can be used to record life cycles of products and devices that contain valuable raw materials. The circular economy product and material data space enables both monitoring of resource and product flows, early detection of market trends for consumer products and cross-references between extraction of materials and their use in products with the aim of assessing the greenhouse effect. This results in a global network, which extends data networks to ultimately track material flows of any color globally.

Depending on the industry and market, intelligent manufacturing enterprises proceed very differently to select their source materials, design, produce, sell, and dispose of their products in different ways too. In this context, digital transformation can offer a sustainable solution for circular economy along the value chain and design, intelligent manufacturing, and recycling processes. It will create new and innovative solutions to circular economy based digital tagging and tracking for generating core data records for a sustainable product and material data space. Based on this approach, the final disposal and recycling of valuable materials used can be achieved. However, this will come up with a new and so far unknown problem area such as cybersecurity, the new and innovative digital solution in the circular economy with their product and material data space.

References

1. C. Sammut, G.J. Webb, *Encyclopedia of Machine Learning and Data Mining* (Springer Nature, New York, 2017)
2. S.S. Schwartz, S. Ben-David, *Understanding Machine Learning* (Cambridge University Press, Cambridge, 2014)
3. T. Kuusisto, R. Kuusisto, Cyber world as a social system, in *Cyber Security: Analytics, Technology and Automation*, ed. by M. Lehto, P. Neittaanmaki, (Springer Publ., Cham, 2015), pp. 31–43
4. T.M. Siebel, *Digital Transformation—Survive and Thrive in an Era of Mass Extinction* (Rosetta Books, New York, 2019)
5. D.P.F. Möller, *Guide to Computing Fundamentals in Cyber-Physical Systems—Concepts, Design Methods, and Applications* (Springer Publ., Cham, 2016). Also available in Chinese Language. ISBN: 978-7-111-59145-0
6. G. Hiemstra, M. Wade, Digital Transformation Closing the Gap Between Now and the Future, CISCO Connected Futures (2017)
7. M.E. Karim, V.V. Proha, Chapter 7: Cyber-physical systems security, in *Applied Cyber-physical Systems*, ed. by S.S. Suh, U.J. Tanik, J.N. Carbone, A. Eroglu, (Springer Nature, New York, 2014), pp. 75–83
8. R.D. Precht, *Hunters, Shepherds, Critics—An Utopia for the Digital Society* (Goldmann, Munich, 2018) (in German)
9. F. Krupp, *How Technology is Driving the Fourth Wave of Environmentalism* (World Economic Forum, Cologny, 2018)

Chapter 2
Introduction to Cybersecurity

2.1 Introduction

The importance of cybersecurity was illustrated by an article in the New York Times in March 2011 describing how researchers were able to hack a car remotely to take control of the car's critical and crucial systems. This was accomplished through the car's embedded communication systems as many of today's cars contain cellular connections and Bluetooth wireless technology [1]. This makes it possible for a hacker, who is able to access from a remote location, to attack various features, such as car door locks and brakes, as well as to track the vehicle's location, eavesdrop on the car's cabin, monitor vehicle data, and many others. In this context complexity increases by integrating communication, computing, and control within today's vehicular systems which plays a dual role with regard to used integrated cyber and physical devices and components. Due to their scale and complexity today's systems and components are vulnerable to a variety of security challenges, intrusions, cyber threat attacks, and malicious attacks besides their designed functionality.

The worldwide availability of the internet allows cyber criminal attackers in today's interconnected digital world launching cyber threat attacks to cyber and physical systems worldwide from anywhere, at any place and at any time, as described in many books, which give a deeper insight into hacking, for instance [2–5]. Therefore, cyber threat attack-related security challenges require effective techniques to combat cyber threat attacks by detecting, preventing, and recovering from cyber threat attacks, for instance malware [6]. However, attacks include both previously known and unknown potential cyber threat attacks. Therefore, cybersecurity becomes a body of knowledge about technologies, processes, and practices developed to protect computers, data sources, networks, and programs from cyber threat attacks, damage, or unauthorized access [7].

D. P. F. Möller, *Cybersecurity in Digital Transformation*, SpringerBriefs on Cyber Security Systems and Networks, https://doi.org/10.1007/978-3-030-60570-4_2

Traditional security approaches focus on using the most relevant resources on the most crucial systems and devices to protect them against the biggest known cyber threat attacks. This may necessitate leaving some less important systems and devices undefended and vulnerable to attack due to less assumed danger or known risks. However, such an approach is insufficient when it comes to digital transformation in public and private organizations which have integrated cyber and physical systems and devices in their environments to launch new business models like *as-a-service* model, *mobility-as-a-service*, and others. This requires an extensive internal transformation across public and private organizations' operational work. Hence, advanced cybersecurity approaches must be developed and used to overcome this gap, because of the growing cyber threat attack landscape. It is well known, that the cyber security threat landscape grew in complexity throughout 2019, with a potent mix of nation state threat actors, cyber-criminal organizations, private sector security providers accelerating the cyber arms race and elevating each other's capabilities, and will no let-up the next years.

2.2 CIA Triad

The term cybersecurity refers to techniques and practices designed to protect data that is stored, transmitted, or used in information systems. Therefore, cybersecurity is one of the cross-cutting issues in digital transformation today, because it is fundamental that authorized messages be delivered at any time <u>and</u> at any place <u>and</u> to the right place <u>and</u> in real time <u>and</u> without any disturbance <u>and</u> without malicious attack. Hence, effective cybersecurity reduces the risk of cyber threat attacks and protects public and private organizations from unauthorized exploitation of cyber and physical devices, networks, systems.

Unfortunately, nowadays online users getting access to data or information on the internet for which they do not have authorization have to be called attackers. In this regard it is important to note, that the ability to hack is a skill, something that involves the manipulation of technology in some specific way, or form. Those who can successfully attack are called attackers or hackers, suggesting that they have the capability to hack. To differentiate between benevolent and malicious hackers, the hacker community introduced the term cracker, to characterize people who engage in criminal or unethical acts using hacking technologies. Thus, this term is meant derisively, suggesting that cracker is different from hacker, and should be treated accordingly [5]. Against this background, information security and cybersecurity is an essential need and has to focus on measures to protect authorized messages from malicious cracker attacks.

The fundamental objective in information security refers to protecting data as well as information systems from unauthorized access, destruction, disclosure, disruption, modification, or usage. Therefore, the three fundamental principles in information security are the principles of Confidentiality, Integrity, and Availability, which are commonly referred to as CIA Triad, a model to guide policies for security within public or private organizations, which also form the main objectives of any security program. The model is also sometimes referred to as AIC Triad (Availability, Integrity and Confidentiality) to avoid confusion with the U.S. Central Intelligence Agency (CIA) term. The elements of the triad are considered as three most crucial components of security. Hence, security of data exchange can be characterized as follows:

- *Confidentiality*: Vital security characteristic in the era of digital transformation. Term is roughly equivalent to privacy. However, it means protecting data from unauthorized access and misuse, for instance by a set of rules that limit access to data. Measures undertaken to ensure confidentiality are designed to prevent sensitive data from reaching the wrong people, making sure that the right people can in fact get it. Federal Code 44 United States Code, Section 3542 defines confidentiality as "preserving authorized restrictions on access and disclosure, including means for protecting personal privacy and proprietary information." This requires a number of access controls and protection as well as ongoing monitoring, testing and training. Data encryption is a common method of ensuring confidentiality. In this regard, user IDs and passwords constitute a standard procedure. Other options include biometric verification, by which a person can uniquely evaluate one or more distinguishing biological traits, as well as security token, which is a small hardware device that an owner carries to authorize access to a network service, and key fobs, which means a small, programmable hardware device that provides access to a physical object, or soft token, a software-based security token, that generates a single-use login PIN. However, to satisfy desired security requirements the solution should include a holistic consideration.
- *Integrity*: Involves maintaining consistency, accuracy, and trustworthiness of data over its entire life cycle. This covers the important topics of data integrity and system integrity. Data integrity is the requirement of data and programs being changed only in a specified and authorized manner, while system integrity refers to the requirement of a system performing its intended function in an unimpaired manner, free from deliberate or inadvertent unauthorized manipulation. Against this background, a deficiency in integrity can allow for modification of data and programs stored on the memory of digital systems used, which can affect the crucial and critical operational functions of the digital systems, without ad hoc detection.
- *Availability*: Information, data and programs are accessible by authorized users when needed and is an essential requirement in the era of digital transformation. This can be ensured by rigorously maintaining all system hardware, immediately performing hardware repairs when needed, and maintaining a correct functioning operating system environment that is free of software conflicts. If crucial and critical operational systems cannot access needed data when required, data, and programs of operational systems are not secure. That availability is a fundamental feature of a successful deployment of digital systems in the era of digital transformation. To prevent data loss, a backup copy may be

stored in a geographically-isolated location, perhaps even in a digital safeguard. Extra security equipment or software such as firewalls and proxy servers can guard against downtime and unreachable data and programs due to malicious activities such as Denial-of-Service attacks, and network intrusions.

The CIA Triad, shown in Fig. 2.1, is a well-known model for the development of security policies used in identifying problem areas, along with necessary solutions in the arena of information, data and system security.

2.3   Cyber Threat Attacks and Cyber Threat Intentions

Cyber threat attacks can be introduced as threats that involve an attempt to obtain, alter, destroy, remove, implement or reveal data without authorized access or permission and thus compromise confidentiality, integrity, and availability. Therefore, cyber threat attacks can show impacts on individuals as well as public and private organizations. Against this background different kinds of cyber threat attacks are available, including but not limited to as (see Sect. 2.6):

- *Active Cyber Threat Attacks*: Attempt to alter attacked system resources or affect their operations. They involve modifications of the data stream or create false statements.
- *Passive Cyber Threat Attacks*: Attempt to make use of data from a system but does not affect attacked system resources. They are more of eavesdropping or monitoring of a transmission. Their goal is to obtain data being transmitted.
- *Botnet Cyber Threat Attack*: Infection by malware of a collection of internet-connected devices that allow hackers to control them. Malware have different forms in which they are used by attackers: Adware, Keylogger, Ransomware, Remote Access, Rootkit, Spyware, Trojan horse, Virus, and Worm. Cyber-criminals use botnets to instigate botnet attacks, which include malicious activities

Fig. 2.1 CIA Triad

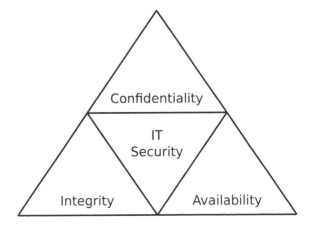

such as credential leaks or credential-stuffing attacks leading to account takeovers, unauthorized access to a device and its connection to a network, data theft to steal data, and Distributed Denial of Service (DDoS) attacks that cause unplanned application downtime (see Sect. 2.6)

- *Brandjacking Cyber Threat Attack*: A phishing scam with activities whereby someone acquires or otherwise assumes online identity of another entity for the purpose of acquiring a person's or business's brand equity. FDA has published a list of tips to avoid phishing scams
- *Clickjacking Cyber Threat Attack*: Attackers use multiple transparent or opaque layers to trick a user clicking on top of a web page. However, on top of that web page, the attacker would have loaded an iframe with an attacked mail account, and lined up an exact 'delete all messages' button directly on top.
- *Inside and Outside Cyber Threat Attack*: Public and private organizations face a growing cyber threat attack from computer criminals committed both inside and outside their organizations. An insider cyber threat attack, also known as insider threat, is a malicious attack perpetrated on a computer system or network by a person with authorized system access of public or private organizations' computer systems or networks. Therefore, an insider is someone with legitimate access to organization resources. Furthermore, an insider is a wholly or partially trusted person. Hence, an insider cyber threat attacker is an individual that has access to organization resources. Moreover, knowing skills of an insider attacker is an additional essential factor posing a cyber threat by the malicious motivation of insider attackers. Unfortunately, there may be less security against insider attacks because many public and private organizations focus on protection from external attacks. An outsider cyber threat attack occurs when an individual attacker or a group of cyber threat attackers seek to gain protected information by infiltrating and taking over profile of a trusted user from outside the organization.
- *Phishing Attack*: Type of social engineering cyber threat attack, used to steal user data, including login credentials and credit card numbers. It is also used to gain foothold in public or private organizations' networks as a part of a larger attack, such as an Advanced Persistent Threat (APT) incident. In this scenario, employees are compromised to bypass security perimeters, distribute malware inside a closed public or private organization environment, or gain privileged access to secured data (see Sect. 2.6)
- *Spamming Attack*: Electronic version of a junk mail, sending unwanted messages, often unsolicited advertising, to a large number of recipients. It is a serious security concern delivering Trojan horses, virus, worms, spyware, and targeted phishing attacks
- And many others.

Two major groups of attackers are of interest, external and internal attackers whereby the internal attacker type can be divided into two groups, employees and insiders. An employee of a company can be an internal hacker who performs exploits within company's networks. If authorized he tries to find vulnerabilities in a company's networks and fixes them. In case the employee is not authorized and attacks to exploit flaws or for some personal gain; he is called a hacker. Insider data breaches can also occur through more accidental means. However, insider threats can also be caused by intentional or malicious attacks from internal employees who are called rogue insiders or criminal attackers. In regard to their behavior profile they may belong to the cracker type.

More in general, an insider cyber threat can be a malicious cyber threat to any public or private organization that comes from someone within the organization, such as employees, former employees, contractors or business associates, who have inside information concerning public or private organizations' security practices, data and computer systems. Insider based cyber threat may involve fraud, theft of confidential or commercially valuable data, theft of intellectual property, or sabotaging of computer systems whereby this behavior profile belongs to the cracker group. Hence, insider cyber threats can be classified into the categories:

- *Malicious Insider Threats*: Someone who takes advantage of their access to inflict maliciously and intentionally abuse legitimate credentials, typically to steal data and harm a public or private organization. This can be executed from frustrated employees, fired employees, unserious contractors or business associates, and others. Also known as "Turncloaks" which have an advantage over other cyber threat attackers because they are familiar with security policies and procedures of a public or private organization, as well as its vulnerabilities [8].
- *Negligent Insider Threats*: Someone who makes unintentional errors or disregard policies, which place their public or private organizations at risk. Also known as "Careless Insider".
- *Infiltrator Threats*: External actors that obtain legitimate wise access credentials based on illegal ways without legal authorization with which they can implement malicious software into organizations systems, known as "Mole".

External data breaches are mostly malicious with the intention of stealing internal data, maliciously modifying or deleting secret data, encrypting data to blackmail money, degrading network performance and affecting organizations operations, as well as providing server failures, intruded by external attackers. These types of cyber threat attacks belong to the group of unknown attacks which an organization did not expect and also do not know when such an attack will happen. This is a dangerous situation for public and private organizations because in such a cyber threat attack valuable time is passed by before a solution is found and implemented to protect the system against such a kind of cyber threat attack.

Internal cyber threat attack intentions by insiders and employees have a broad range of cyber threat attack possibilities and are among the most difficult ones to be detected and prevented. Insiders and employees gain systems or network access as well as systems or network knowledge. Hence, they have physical access to critical and crucial areas in public or private organizations and are able to take access with less effort. They may become motivated by revenge or entitlement when employment is terminated and they often take customer or company data with them when moving to a competitor company, or take revenge on their company by sending proprietary data to competitors over the internet. Another type of insiders who may crack the organization's system to cause damage is often angered employees who have been fired from their jobs and have computer skills. They can, for instance, sabotage the company's network planting logic bombs that damage after the employees leave.

Thus, intrusion detection systems (IDSs) and intrusion prevention systems (IPSs) are essential as they can be used to support detection and prevention of cyber threat attack intentions (see Chap. 4). However, the question to be answered is how to identify a potential insider or employee intruder as attacker type hacker. One way of doing so is by using insider or employee data, based on account identification which can help tracking back by comparing insider or employee account data with attack data to identify identical signatures.

Another important cyber threat attack type is password guessing. This is one of the most common cyber threat attack types by insiders or employees. Here, an insider or employee attacker knows the login details and then attempts to guess the password for a security breach for a successful login.

Finally, this raises the question: "What personal traits do these insiders or employees have?" After analyzing a pool of cracking cases provided by computer crime investigators, for instance prosecutors and security specialists, the researchers conclude that insider or employee computer criminals tend to be [9]:

- Introverted individuals who admit to being more comfortable solving cognitive problems than interacting with others in the workplace,
- More dependent on online interactions than on face-to-face interactions,
- Ethically flexible individuals who can easily justify ethical violations,
- Of the opinion that they are somehow special and thus deserving of special privileges, or
- Lacking in empathy and thus seeming not to reflect on the impact their behaviors have on others or on the company or public or private organizations.

2.4 Security Risk, Likelihood, and Consequence Level

From Sect. 2.1 it can be seen that defending potential cyber security breaches is one of the most important tasks for cybersecurity. Therefore, today's engineered digital information and communication systems and networks must be designed by security issues. They are built from and depend upon the seamless integration of computational algorithms and physical objects composed of sets of wireless networked components, including sensors, actuators, control processing elements, and communication devices, which are critical and crucial against cyber threat attacks. Hence, using these smart and highly reliable systems and networks, one must carefully consider possible vulnerabilities of these systems and networks, which may result in potential security problems. In fact, concerns with security of systems and networks include malicious attempts through cyber threat attacks. These types of cyber threat attacks affect a large group of mission-critical systems and networks, which could result in denying essential services, stealing sensitive data, causing various types of damage, and many others. Thus, cybersecurity, from a general perspective, deals with risk analysis. Once a major risk for an unauthorized intrusion has been identified, an analysis is carried out to determine the likelihood (probability) of the risk occurring and the consequence (impact) of that risk should it occur, which is called risk quantification. However, understanding the impact and potential consequences of risks in systems or components of systems requires a solid understanding of risk and vulnerability types to decide about likelihood and impact. But many risk models suffer from vague, non-qualified output based on partial information or on unfounded assumptions. Hence, threat intelligence provides context that helps risk models make defined risk measurement and be more transparent about their assumptions, variables, and outcomes. A simple classification scheme can be derived for distinct risk levels, as shown in Table 2.1.

A detailed analysis of well-known and documented cyber threat attacks in today's critical and crucial systems with their environments, such as Stuxnet [10] or Flame [11], might reveal identifiable vulnerabilities interactions or dependency signatures. In case of such recognizable signatures, further investigation has to be done to identify specifications, severity, impact, counter-measures, and if possible, development of a methodological approach to reveal them before an incident happens. With this clear understanding of interactions and dependencies of cyber threat vulnerabilities, a cybersecurity framework can be developed in terms of essential assets. However, if different types of risks exist along the supply chain value this can be assumed as associated shared value chain cyber risks by data breaches, or cyber threat attacks. To avoid cyber threat attacks, cybersecurity methodological approaches must extraordinarily evolve to reveal before cyber incidents happens. Against this background, a methodological approach should use qualitative values for likelihood levels, described as frequency values, meaning how easy it is to exploit cyber threat attacks, as shown in Table 2.2. Table 2.3 classifies consequence levels against appropriate impact of cyber threat attacks.

Table 2.1 Risk levels and their possible impact

Risk level	Impact
High	Not acceptable risk level: Identified threats be classified to defend them successfully and in real-time
Medium	Moderate risk level: Identified threats must be monitored with the consideration of whether necessary measures must be taken to defend a potential incident
Low	Acceptable risk level: Identified threats must be observed to discover changes that could increase the risk level

Table 2.2 Likelihood levels and possible threat frequency

Likelihood	Frequency of threat case or time of threat
Highly Likely	Threats occur very often, for example more than every tenth access. This means that threats occur more frequently, i.e. in 10% of threat cases or time required for decision making
Likely	Threats occur quite often, for example, less than every tenth access, which means that threats occur frequently, i.e. between 1% and 10% of threat cases or time required for decision making
Possibly	Threats may occur, for example, less than every 1 access, which means less frequently, i.e. between 0.1% and 1% of threat cases or time required for decision making
Unlikely	Threats occur rarely, for example, lower than every 0.1 access, which means less frequently, i.e. less than 0.1% of threat cases or time required for decision making

Table 2.3 Consequence levels and possible cyber threat impact

Consequence level	Cyber threat impact
Catastrophic	Cyber threat can be classified as severe level risk with successful devastating impact and loss of trust
Severe	Cyber threat can be classified as a serious level risk with a serious level of successful impact and loss of trust
Moderate	Cyber threat can be classified as a low-level risk with a low level of successful impact that may influence trust
Small	Cyber threat can be classified as one with no strong/dangerous risk

2.5   Risk Matrix

There are several types of risk and several ways to quantify risk for analytical assessments. The standard deviation is a common metric associated with risk, providing a measure of the volatility of a value in comparison to its past average. However, more in general the risk value level for cyber threat attacks can be calculated as the product of consequences and likelihood values, as illustrated with more detailed segmentation of likelihoods in a two-dimensional risk matrix in Table 2.4. The shading of the matrix visualizes different risk levels. Based on the acceptance criteria in Table 2.1, risk level "High" is decided as unacceptable severe level risk. Any cyber threat attack obtaining this risk level must be treated in order to have its risk reduced to an acceptable level. Only risk level "Catastrophic" describes an unacceptable catastrophic risk that must be treated immediately to minimize a malicious impact. Thus, the respective conditions for cybersecurity require a risk level analysis. The risk level itself can be determined in the one hand as sum of all individual cyber threat attacks in a given reference time window t_R or in the other hand as a product of average likelihood levels of damage risk frequency.

As a rule, security is achieved when an existing risk level R_{LE} is less than a reasonable risk level R_{LR}, for instance, an acceptable marginal risk, as defined in Table 2.2. Hence, security (S) can be described as a relation between actual risk level R_{LA} and a minimal risk level R_{LM} as follows:

Table 2.4 Risk matrix as a function of risk levels

Likelihood	Consequence				
	Catastrophic	Severe	Moderate	Small	Insignificant
Highly Likely	Catastrophic	High	Medium	Medium	Low
Likely	High	High	Medium	Medium	Low
Possible	Medium	Medium	Medium	Low	Low
Unlikely	Medium	Medium	Low	Low	Low
Rare	Low	Low	Low	Low	Low

$$S \quad = \quad R_{LA} < R_{LM}$$

The minimal risk level (R_{LM}) can be considered as the accepted marginal risk level (R_{LAM}). However, cyber attackers may perform cyber threat attacks (C_{TA}) which describe the actual executed risk level (R_{LAE}) that is introduced as an element of a set

$$a_R : C_{TA} \in a_R$$

whereby a_R represents the attack risk.

Let's assume that a cyber attacker has a minimal number of resources r to support his cyber-attack effort. Then cyber threat attacks (C_{TA}) that a cyber attacker (C_A) is able to perform can be summarized by subset

$$a_R C_A C_{TA} \subseteq a_R.$$

If a cyber threat attack occurs, its success can be indicated through the degree $\mu \in [0, 1]$. In case a successful cyber threat attack happens, $\mu = 1$. In case an unsuccessful cyber threat attack happens, $\mu = 0$. There are several methods evaluating the threshold μ of cyber threat attack risk levels which may result in an alarm. However, before an alarm is activated, the person or organization is alerted, to warn of a danger or problem, typically with the intention of observing it to discover intrinsic changes that could result in an increasing or decreasing risk level.

The next possibility in an alarm and warning sequence at lowest level is alerting. Alerting means that the risk level is at a likelihood level of unlikely, as shown in Table 2.2, and typically alerting has the possibility to neglect it or deal with it which is considered as accepted marginal risk level (R_{LAM}). Let's assume the targeted system consists of several sub-systems, and the different sub-systems have to be analyzed as described above. This analysis may result in the probability of sub-system cyber threat attack risk, as shown in Table 2.5.

Furthermore, vulnerability through a cyber threat attack can be modeled by attackers' goals and beliefs. But it is not sufficient only knowing what the attackers' believe and desire, it is also important to know how strong these attitudes are. However, beliefs vary in strength, which has been proved not only in decision theory, but also in artificial intelligence, statistics, and many other areas. Thus, the theory of belief functions provides a way using mathematical probability in subjective judgement.

Table 2.5 Risk matrices as a function of system or sub-system risk levels

Likelihood	Consequence				
	Catastrophic	Severe	Moderate	Small	Insignificant
Highly Likely	Catastrophic	High	Medium	Medium	Low
Likely	High	High	Medium	Medium	Low
Possible	Medium	Medium	Medium	Low	**Low** R_{LAM} system 1 has a warning
Unlikely	Medium	Medium	Low	**Low** R_{LAM} system 1 component 4 has a warning	**Low** R_{LAM} system 1 component 3 has an alert
Rare	Low	Low	Low	**Low** R_{LAM} system 2 has an alert	**Low** R_{LAM} system 5 has an alert

In this regard it is a generalization of the Bayesian theory of subjective probability. Using the Bayesian theory to quantify judgments about a question, one must assign probabilities to the possible answers to that question. Hence, the theory of belief functions is more flexible; it allows deriving degrees of belief for a question from probabilities for a related question. These degrees of belief may or may not have the mathematical properties of probabilities; how much they differ from probabilities will depend on how closely the two questions are related [12]. With this in mind, Bayesian models derive from the assumption that rational degrees of belief satisfy the mathematical conditions of a probability function. Among other things, this means that the credence assigned to a status in a decision problem must add up to 1, which can be introduced as a degree of belief (*DoB*) or credence (*Cr*), which obey the probability calculus. The idea beyond is that belief comes in degrees based on the observation being more certain of some things than of others. Hence, a degree of belief is a rational and objective to give available evidence which may also rise to contradictory results.

For a simple example two questions, q_1 and q_2, are given with probabilities for q_1 and derived *DoB* for q_2. Let's assume that q_1 and q_2 had only two possible answers: yes and no, then no formal notation is required. Let's now assume that more possible answers are possible. In this context a notation is required for each question's set of possible answers, as well as a notation for the possibilities for q_1 and the *DoB* for q_2, representing the constraints that an answer to q_1 may put on the answer to q_2. Let's also assume that one of the answers is correct, but it is not known which one. Now the set of answers S_A can be assumed as a frame for q_1, the question for which probabilities exist, and T can be a frame for q_2, the question of interest. Let $p(s)$ be the probability of element s of S_A. For $p(s)$ and a subset A of T the degree of belief $DoB(A)$ is given which means that A contains the correct answer to q_2.

Against this background, an approach for solving decision problems evaluates each form by the weighted average of the utility of all possible outcomes, weighted by the likelihood of the relevant status, as given by the attacker's *DoB* or *Cr*. For a simple case study, averages can be calculated as follows for the numbers x_1, x_2, ..., x_n, and then the average of the numbers is

$$\frac{x_1 + x_2 + \ldots + x_N}{N} = \frac{1}{N}x_1 + \frac{1}{N}x_2 + \ldots + \frac{1}{N}x_N.$$

In this simple case each number has the same weighted average; but the weight can also be different for different numbers. Let's assume a cyber threat attack leads to outcomes

$$O_1, \ldots, O_N$$

for system status

$$S_{S1}, \ldots, S_{SN}$$

Let's also assume $DoB(s)$ denote the cyber threat attack's degree of belief in Ss_1, $DoB(Ss_2)$ the degree of belief in Ss_2, and finally $DoB(Ss_N)$ the degree of belief in Ss_N. Let $E(O_1)$ denote the expected outcome of O_1 for the first identified cyber threat attack, $E(O_2)$ the expected outcome of O_2 for the second identified cyber threat attack, and finally $E(O_N)$ the expected outcome for O_N for the last identified cyber threat attack. Then the expected cyber threat attack vector (EC_{TA}) of identified cyber threat attacks can be defined as follows:

$$EC_{TA} = D_0 B\left(S_{S1}\right) E\left(O_1\right) + D_0 B\left(S_{S2}\right) E\left(O_2\right), \ldots, D_0 B\left(S_{SN}\right) E\left(O_{2N}\right)$$

which can also be written in summarizing form:

$$EC_{TA} = \sum_{i=1}^{N} D_0 B\left(S_{Si}\right) E\left(O_i\right)$$

From practical experience, subjectivism gives the same probabilities to frequently repeated events, where the probability is characterized as relative frequency of incidents in the long run, as shown for instance in Table 2.2.

2.6   Cyber Threat Attack Types

Risky end-user behavior in public and private organizations around the world impacting implications like immediate infection by ransomware, or a cyber threat that lies in wait as an incident of credential compromise, is an important issue. What has to be aware is that users' personal cybersecurity habits carry over into worktime, and that, often, info security teams are overestimating end users' understanding of fundamental cybersecurity practices or more general knowledge [13]. Therefore, a much deeper knowledge is required to identify cyber threat attacks and defend the system or network. This requires on the one hand to identify cyber threat attacks and on the other hand to characterize attackers' intentions based on the CIA Triad (see Sect. 2.2), and finally to determine the likelihood versus the cyber threat attack impact (see Tables 2.3 and 2.4) for further cybersecurity action. More in general, the most common cyber threat attacks against public and private organizations are phishing attacks, negligent and malicious insider attacks, advanced persistent threats, cyber threat attacks, zero day exploits, attacking known software vulnerabilities, denial of service attacks, brute force attacks, and many others. However, cyber threat attacks are typically composed of a single or a combination of different types of options such as:

- *Advanced Persistent Threat (APT)*: Network cyber threat attack in which unauthorized persons access a network and stay there undetected in the long term. The primary intent of an APT is to steal data, disrupt business operation and damage infrastructures. APT attackers coordinate their activities with the security measures of their targeted private or public organization and often attack them several times. APT groups often receive instruction and support from governments or government agencies.
- *Botnet*: Sets of computers that are under control of a malicious controller used without the knowledge of their owners to send files (including spam and viruses) over the Internet to other computers. Every element of this system is called a bot. Most of the affected systems are private computers. Bot computers can work in distant directions without the owner's knowledge. However, there are a few clues that can indicate a possible botnet attack: slow computing speed, high CPU usage as well as unnecessary sudden pop-ups. As an example, Emotet is a malicious botnet software type.
- *Brute-Force-Attack*: Trial-and-error method to obtain information such as a user password or personal identification number (PIN). It is based on software which autonomously generates a large number of consecutive guesses as to the value of the desired data. However, this attack needs time to run to provide anything usable. This attack is used by cyber-criminal attackers to crack encrypted data, but also by security analysts to test an organization's network security. Therefore, brute force attacks can be defended by increasing password length or increase password complexity which increases time to brute force crack.
- *Cross-site Scripting Attack*: Represent attacks which use third-party web resources to run scripts in the attacked web browser or scriptable application. The cross-site scripting attack type attacker injects a payload with malicious JavaScript, or VB Script, or Active X, or Flash, into a website's database. When a website user requests a page from the website, the website transmits the page with the attacker's payload as part of the HTML body to the user's browser, which executes the malicious script. Cross-site scripting attacks can also exploit vulnerabilities that can enable an attacker to steal cookies, log key strokes, capture screenshots, discover and collect network information and remotely access and control the attacked user's computer system or network.

- *Data Destruction*: Process destroying data stored on electronic/digital media so that it is completely unreadable and cannot be accessed or used anymore. However, it should be noted, that data destruction is not the same as destroying the media on which data is stored which is a physical destruction.
- *Data Manipulation*: Cyber threat attack form of an indirect type of sabotage by altering data to indirectly compromise a project that decisions are based on bad data which have the potential to cause a great damage later.
- *Denial of Service (DoS) Attack*: Cyber threat attack form that shuts down a machine or network, making it inaccessible to its intended user(s). The two general DoS attack methods are accomplished by flooding the target with traffic for the server to buffer, causing them to slow down and eventually stop, or sending it information that triggers a crash.
- *Distributed Denial of Service (DDoS)*: Cyber threat attack to disrupt normal traffic of a targeted server, service or network by overwhelming the target or its surrounding infrastructure with heavy internet traffic.
- *Drive-By Attack*: Refers to drive-by-download attacks, a common method of spreading malware. Hackers look for insecure websites and embed a malicious script into HTTP or PHP code on one of the pages. The script might install malware directly on the computer of a user who visits the website or it might redirect the user to a website controlled by the hacker(s). Drive-by downloads can occur when visiting a website or viewing an email message or a pop-up window. Unlike many other types of cyber threat attacks, a drive-by-download attack doesn't rely on a user to do anything to actively enable the attack. The attacked user doesn't have to click a download button or open a malicious email attachment to become infected. A drive-by download attack can take advantage of an app, operating system or web browser that contains security flaws due to unsuccessful updates or lack of updates. It is an unintended download of a malicious code on a computer or another device. Unlike many other types of cyber threat attacks, a drive-by download does not rely on the user to do anything to actively enable the attack.
- *Eavesdropping Attack*: This attack type occurs through the interception of network traffic. Attackers can obtain passwords, credit card numbers and other confidential information the users send over the network. Eavesdropping can be passive or active [14]. Detecting passive eavesdropping attacks is often more important than spotting active ones, since active attacks require the attacker to gain knowledge of the friendly units by conducting passive eavesdropping before.
- *Intellectual Property Theft*: Cyber threat attack through digital technologies and internet file sharing networks, robbing ideas, inventions, and creative expressions which can include everything from trade secrets and proprietary products and parts to movies, music, and software.
- *Man in the Middle Attack*: Cyber threat attack when a perpetrator positions himself in traffic between a targeted user and an application, either to eavesdrop or to impersonate one of the parties, making it appear as if a normal exchange of information is underway to steal personal information. Critical to this attack is that the targeted user is not aware of the man in the middle.

- *Malvertising*: Cyber threat attack in which perpetrators inject malicious code into legitimate online advertising networks. The code typically redirects users to malicious websites.
- *Malware Attack*: Refers to malicious software that is installed in a computer system or network without consent. It can attach itself to legitimate code, propagate and replicate itself across the internet.
- *Negligent and Malicious Insiders*: Employees, associates and/or affiliates who have legitimate access to an IT system in organizations. Maliciously focused on what assets are at risk of leaving the organization through the IT environment as well as threats entering the organization through the same means.
- *Password Attack*: Attack mechanism to authenticate users to an information system, obtaining passwords is a common and effective attack approach. Access to a person's password can be obtained by looking around the person's desk, "sniffing" the connection to the network to acquire unencrypted passwords, using social engineering, gaining access to a password database or outright guessing. The last approach can be done in either a random or systematic manner.
- *Phishing*: Form of a fraud in which a cyber threat attacker masquerades as a reputable entity or person in email or other communication channels. The cyber threat attacker uses phishing emails to distribute malicious links or attachment that can perform a variety of functions, including the extraction of login credentials or account information from attacked computer systems or networks to gain sensitive, confidential information such as usernames, passwords, network credentials, and more, by posing as a legitimate individual or institution (see Sect. 2.3).
- *Pre-phishing Attack*: Cyber threat attack that tries to uncover names, job titles and email addresses of potential victims, as well as information about their colleagues and the names of key employees in their organizations. This information can then be used to craft a believable email. Targeted attacks, including those carried out by APT groups; typically begin with a phishing email containing a malicious link or attachment.
- *Ransomware*: Type of malware in which the data on a targeted computer system or user's files is locked, typically by encryption, and payment is demanded before the ransomed data is decrypted and access is returned to the targeted user. In recent years, over 50 different ransomware variants with names such as Cryptolocker, CryptoWall, KeRanger, Locky, TeslaCrypt, and many others, have been spotted on the Internet. As soon as activated by an unsuspecting user, the ransomware contacts a control server, which sends it a randomly generated Advanced Encryption Standard (AES) key. AES is a symmetric key encryption cipher which means that the same key used to encrypt the data is used to decrypt it. Hence, AES key is used to encrypt important files on the local hard drive as well as drives in the network and in the connected cloud. From this point on, the data is completely under the control of the cyber-criminal attacker, the hacker, who can now request an immediate ransom to restore the files or not to publish them. Public and private organizations can make use of virtualization, cooperative mobility management, a file synchronization to protect their computer systems, tablets, smartphones and other devices from being infected with

ransomware. Moreover, the public or organizations' data can be quickly restored in case of a successful ransomware attack. Furthermore, avoiding ransomware payment if a computer system is hit by a ransomware attack can be done by air gapping as part of 3-2-1 backup strategy, which means having an offline copy of the latest backup to use for recovery and not to pay the ransom.

- *Rogue Software*: Type of internet fraud using computer malware to trick users into revealing financial and social account details or paying for bogus products. Rogue software misleads users into believing there is no virus on the computer and aims to convince them to pay for a fake malware removal tool that actually installs malware on the computer.
- *Spyware*: Software installed on a computing device without end user's knowledge which can violate end user's privacy and has the potential to be abused. The software can be difficult to detect; often, the first indication a user has that a computing device has been infected with spyware is a noticeable reduction in processor or network connection speeds and in the case of mobile devices' data usage and battery life
- *Standard Query Language (SQL) Injection Attack*: Injection with database-driven websites. It occurs when a malefactor executes a SQL query to the database via the input data from the client to server SQL commands are inserted into data-plane input in order to run predefined SQL commands, Hence, a successful SQL injection exploit can read sensitive data from the database, modify (insert, update or delete) database data, execute administration operations (such as shutdown) on the database, recover the content of a given file, and, in some cases, issue commands to the operating system.
- *Wiper Attacks*: Malware with the sole intention of destroying systems and/or data, usually causing great financial and/or reputation damage.
- *Zero-Day-Exploit*: Cyber threat attack that occurs on the same day a weakness is discovered in software. Weakness is exploited before a security patch to fix the flaw becomes available from the software creator. If it happens, there is little protection against a cyber threat attack because the software flaw is new.
- And many others.

Against this background, cybersecurity brings a major concern with regard to cyber threat attacks to systems or networks in the public and private domains. Thus, cybersecurity will remain the paramount for security risks.

References

1. R.Y. Fahmida, GM's OnStar, Ford Sync. MP3 Bluetooth attack vectors for cars, eweek.com (2011)
2. P. Kim, *The Hacker Playbook* (Secure Planet Publ., North Charleston, SC, 2014)
3. K. Mitnick, W.L. Simon, S. Wozniak, *Ghost in the Wires* (Little Brown and Company, New York, 2011)
4. M. Goodman, *Future Crimes* (Penguin Random House, New York, 2016)

5. T.J. Holt, B.H. Schell, *Hackers and Hacking* (ABC-CLIO Press, Santa Barbara, CA, 2013)
6. M. Sikorski, A. Honig, *Practical Malware Analysis* (No Starch Press, San Francisco, CA, 2012)
7. D.P.F. Möller, R.E. Haas, *Guide to Automotive Connectivity and Cybersecurity—Trends, Technologies, Applications, and Innovations* (Springer Nature, Cham, 2019)
8. https://www.imperva.com/learn/application-security/insider-threats/
9. E. Shaw, K.G. Ruby, J.M. Post, The insider threat to information systems—the psychology of the dangerous insider. Security Awareness Bull. **2**, 1–10 (1998)
10. R. Langner, Stuxnet: dissecting a cyberwarfare weapon. IEEE Security Privacy **9**(3), 49–51 (2011)
11. B. Miller, D. Rowe, A survey SCADA of and critical infrastructure incidents, in *Proceedings of the 1st ACM Annual Conference on Research in Information Technology*, 2012, pp. 51–56
12. G. Shafer, Perspectives on the theory and practice of belief functions. Int. J. Approx. Reason. **4**, 323–362 (1990)
13. https://info.wombatsecurity.com/hubfs/WombatProofpoint-UserRiskSurveyReport2018_US.pdf
14. https://blog.netwrix.com/2018/05/15/top-10-most-common-types-of-cyber-attacks/#Malware%20attack

Chapter 3
Threat Intelligence

3.1 Introduction

Cyber threats' intent is to inflict harm. Hence, cyber threat perception can be described as an estimated capability and estimated intention to vulnerability and opportunity to execute the threat. Opportunity incorporates understanding of both the cyber threat actor and the cyber threat defender and can be defined as a favorable time or opportunity for a cyber threat actor in relation to a cyber threat defender. Thus, threat intelligence is the information one has to deal with to understand the cyber threats that target or will target the respective resources. This information is essential to identify and to prevent malicious software intrusion in valuable resources, because cyber threats include different forms of attacks and techniques as well as malware and physical threats. Some of the major cyber threat agents in cyberspace are corporations, cybercriminals, employees, hacktivists, nation states, and terrorists [1, 2]. Hence, it is a high level need to educate employees in cybersecurity [3] and about motivational factors such as cyber activism, cyber-crime, cyber espionage, cyber terrorism, and cyber warfare. With such a topology, motives can be reduced to their very essence: anarchy, destruction, egoism, money, and power [2].

Traditional approaches for cybersecurity, that focus only inward on understanding and addressing risks, vulnerabilities, weaknesses and configurations are necessary but not sufficient as they should be. The reason is that mobile technology and the so called social media offer entirely new vistas for modern cyber threat attacks for instance cyber swarming which has assorted different forms of motives [2]. Hence, the effective defense against current and future cyber threat attacks also requires an outward focus, on understanding an adversary's behavior, capability, and intent. Thus, threat intelligence is assumed to provide the way for gathering data about who is attacking, what motivation and capabilities are available at the cyber threat attacker's side, and what kind of indicators can be identified that will help to make decisions for developing a successful cybersecurity strategy. Against this

D. P. F. Möller, *Cybersecurity in Digital Transformation*, SpringerBriefs on Cyber Security Systems and Networks, https://doi.org/10.1007/978-3-030-60570-4_3

background, a balanced understanding of both cyber threat attackers' motivational factors and cyber threat defending team knowledge that allows enough understanding about the nature of cyber threats private and public organizations face, to come up with intelligent threat defense decisions. Alongside commoditized cyber threats, today advanced capabilities that were rare in the past are now commonplace. The reason for this is that today's cyber world is a global and multidimensional information and communication technology network, into which cyber threat attackers can connect via fixed, remote or mobile data nodes, and virtually move within. However, the data used can be unstructured as well as from different disparate sources to be used for Indicators of Compromise (IoC) and Tactics, Techniques, and Procedures (TTP) of cyber threat attackers. Indicators of Compromise are the evidence that a cyber threat attack has taken place. The IoC give valuable information about what has happened but can also be used to prepare for the future and prevent against similar cyber threat attacks. TTP describe an approach of analyzing an Advanced Persistent Threat (APT) operation, a cyber threat attack scenario showing a series of steps, ending with the cyber threat attacker having an established foothold in the attacked computer system or network. APT attackers are typically assumed to be nation states but the same behaviors can also be exhibited by those engaged in conducting cyber-crime, financial threats, industrial espionage, hacktivism, and terrorism, whereby cyber-crime covers traditional forms such as fraud or forgery, publication of illegal over electronic media, and attacks against information systems, denial of service or hacking [2, 4]. Cyber-terrorism is a new form of cyber threat attacks against mission critical and crucial systems. Hence, cyber threat intelligence seeks to understand and characterize narratives like: <u>what</u> sort of cyber threat attack actions have occurred or are likely to occur; <u>how</u> can these actions be detected and recognized; <u>how</u> can they be mitigated; <u>who</u> are the relevant cyber threat attackers; <u>what</u> are they trying to achieve; <u>what</u> are their capabilities in form of TTP; do they have leveraged over time or likely to leverage in the future; <u>what</u> sort of vulnerabilities, misconfigurations, or weaknesses are they likely to target; <u>what</u> actions have they taken in the past [4].

3.2   Threat Intelligence Methodological Approach

Threat intelligence is evidence-based knowledge, including context, mechanisms, indicators, implications and actionable advice, about an existing or emerging menace or hazard to assets that can be used to inform decisions regarding the subject's response to that menace or hazard [5]. Thus, threat intelligence is an essential part of any proactive cybersecurity strategy gathering data prior to a malicious cyber threat actor interacting with the defended computer system or network. However, defending cyber threat attacks is also sharing of data on the latest observed cyber threat attacks. Such data may be collected by a third party, shared between public and private organizations, or shared between a group of public and private organizations. But an inherent systemic issue exists that the vast majority of it is inherently reactive. Rapidly changing cyber threat attacks however, often cause cybersecurity information standards to often become available too late. Cyber threat attackers who specialize in finding exploits and developing malware platforms continue to improve their craft especially with regard to how stealthily their malware infects and operates. Malware is specifically designed software to disrupt, damage, or gain unauthorized access to a computer system. Computer system or network vulnerability allows the malware to perform actions on computer systems or networks that should not be permitted, such as running arbitrary code. Such malicious actions can impact confidentiality, integrity, or availability of computer systems or networks. As reported in [6], zero-day-malware may exist in the wild for over 300 days before identification. Hence, moving toward more proactive threat intelligence is required, for instance based on forward learning to capture the essence of the ever evolving nature of cyber threat attacks. Therefore, threat intelligence is widely being seen as a domain of data scientists, a profession investigating complex data from the business perspective. Data scientists make predictions helping public and private organizations to take accurate decisions with regard to their cybersecurity. Thus, they must have a solid foundation in computer science and network communication, data analytics, applied mathematics such as statistics and stochastics, as well as modeling skills and business informatics, to deal with the different business and information technology requirements. Hence, data scientists are often responsible to identify the relevant cyber threat attack problems to be solved in their respective public and private organization to reveal value in cybersecurity to the organization after resolving it. This is often treated as a separate option in the cybersecurity framework which depends on data collection, data processing and data analytics. Hence, threat intelligence can be broken down in subcategories:

- *Strategic Threat Category*: Non-technical risk-based approach. Allow insights into mission critical and crucial risk areas helping managers and operators in public and private organizations to act in conjunction with action lines, signatures in cyber threat actor's tactics and targets as well as in geopolitical trends. In [7], the usefulness of the Violent Extremist Risk Assessment (VERA) method is applied to five groups of cyber-criminal attackers to determine whether it is more applicable to cyber threat actors who work alone or as part of a group. The method enables a systematic and structured assessment of actors with regard to their risk of violence. Thus, a threat intelligence tool was developed on the basis of structured professional judgment method, which takes individual characteristics of a case into account and finally allows various possible courses of the case included in the planning of measures. The strategic cyber threat attack category has tremendous value for business decision-making but is only one aspect out of broader threat intelligence options.
- *Tactical Threat Category*: Outlines Tactics, Techniques, and Procedures (TTP) of cyber threat attacks to help cyber threat response teams to understand how their public and private organizations might be attacked, and what is the best way defending against or mitigate these attacks.

- *Operational Threat Category*: Gaining knowledge about cyber threat incidents to help cyber threat incident response teams to understand the nature, intention, and timing of potential cyber threat incidents.
- *Technical Intelligence*: Refers to technical threat indicators, e.g., malware hashes, a commonly-shared form of threat intelligence practiced by sharing host-based indicators for malicious code, which are often file names and hashes, C to Integer Program (C2IP) addresses, and others. A hash of a file, means a computed cryptographic checksum of the file. In this regard hashes are outputs of a hashing algorithm, that essentially aim to generate a unique, fixed-length string, the hash value, for any given piece of data. Thus, hash value is of great help to cybersecurity research, malware defense teams, sharing Indicators of Compromise (IoC), and others. Using hash values, cybersecurity researchers can reference malware samples and share them with others through malware repositories like VirusTotal, VirusBay, Malpedia, and MalShare.

An essential issue of these subcategories is how to gain the respective knowledge for cyber threat response teams, cyber threat defenders or cyber threat incident response teams that they are able to answer the question "Who will potentially benefit from cyber threat attacks?" For answering such kind of questions, sources of data are essential. In this context pre-reconnaissance in threat intelligence, refers to data gathered before malicious threats, interacts with defended computer systems or networks in public and private organizations, detecting vulnerability or a cyber threat attack and possibly repelling it. Detecting hostile cyber threat attacks depends on the number and type of appropriate actions, obtained from publicly available data, found for instance in the

- *National Vulnerability Database (NVD)*: U.S. government repository of standards-based vulnerability management data represented using the Security Content Automation Protocol (SCAP). This data enables automation of vulnerability management, security measurement, and compliance. NVD includes databases of security checklist references, security-related software flaws, misconfigurations, product names, and impact metrics [8].
- *Common Vulnerabilities and Exposures (CVE) Database*: List of entries, description, and public reference for publicity known cybersecurity vulnerabilities. Entries are including in the NVD [8].
- *Relationships*: CVE List feeds NVD, which then builds upon the data included in CVE entries to provide enhanced data for each entry such as fixed information, severity scores, and impact ratings. As part of its enhanced data, NVD also provides advanced searching features such as by OS; by vendor name, product name, and/or version number; and by vulnerability type, severity, related exploit range, and impact [9].

These databases are sponsored by the U.S. Department of Homeland Security, the U.S. Office of Cybersecurity and Communications, and the Computer Emergency Readiness Team (CERT) helping understanding the severity of the current security cyber threat landscape. Recent years have seen the adoption of open standard languages and protocols. The MITRE Corporation has published a document with the title "Standardized Cyber Threat Intelligence Information with the Structured Threat Information eXpression (STIX™)" [8], which reflects ongoing efforts to create, evolve, and refine a community-based development sharing and structuring cyber threat attack data. MITRE Corporation, a non-profit organization for the operation of research institutes on behalf of the United States Government, was created by separation from the Massachusetts Institute of Technology (MIT). The Structured Threat Information eXpression (STIX™) is an evolving, collaborative community-driven activity, used to exchange Cyber Treat Intelligence (CTI). It enables public and private organizations to share CTI with one another in a consistent and machine readable form, allowing security communities to better understand what computer-based attacks they are most likely to see and to anticipate or respond to those cyber threat attacks faster and more effectively. In this context STIX™ is designed to improve many different capabilities, such as collaborative threat analysis, automated threat exchange, automated detection and response [10]. Hence, STIX™ defined and developed a language and serialization format to represent structured cyber threat information to convey the full range of cyber threat information and strives to be fully expressive, flexible, and extensible, automatable, and as human-readable as possible. STIX™ is built upon feedback and active participation from public and private organizations and experts across a broad spectrum of industry, academia, and government. This includes consumers and producers of cyber threat attack information in security operations centers, Computer Emergency Response Teams (CERT), cyber threat intelligence activists, security executives and decision makers, as well as numerous currently active data sharing groups, with a diverse set of sharing models. CERT is a group of experts who respond to cybersecurity incidents. These teams deal with the evolution of malware, viruses, and other cyber threat attacks. Against this background the Transport Protocol Automated eXchange of Indicator Information (TAXII™) allow the sharing of security data. These open standards allow the secure sharing of cyber threat information. However, due to the advancement in cyber threat attacks, there is an urgent need for new and outward-looking collaborative approaches to cybersecurity defense. Therefore, threat intelligence and cyber threat information sharing are on the cutting-edge of novel approaches with high potential shifting the balance of power between cyber threat attackers and cyber threat defenders. However, a core requirement for maturing effective threat intelligence and cyber threat attack information sharing is at the availability of an open-standardized structured representation for cyber threat attack information. In this regard, STIX™ is a community-driven effort to provide such a representation adhering to guiding principles to maximize expressivity, flexibility, and extensibility. All parties interested in becoming part of the collaborative community discussing, developing, refining, using and supporting STIX™ are welcome. (http://stix.mitre.org/).

Furthermore, developing a Threat Intelligence Model (TIM) will be another helpful approach in analyzing cyber threats' potential intention. This will allow identifying what kind of data can be collected investigating single as well as multiple cybersecurity incidents, to gain threat intelligence. In this context TIM can be developed based on capabilities referred to as preventive and detective capabilities, forming some kind of profiling approach. As described in [11], this type of model contains the following elements at the detective level:

- *Identifying Identity of a Threat Actor*: This refers to the name of a person, a public and private organization, a nation state, and others. Sometimes, the identity can only be linked to other cyber threat attacks without actual attribution or even location of their operations. However, it is important to connect multiple cyber threat attacks to the same threat actor in order to determine any strategy, Tactics, Techniques, and Procedures (TTPs) expected to be used.
- *Motivation*: Driving force that enables actions in the pursuit of specific goals. Motivation can be derived from the benefits achieving a goal. Cyber threat attackers' goals may change, but the motivation stays the same, most of the times. Knowing a threat actor's motivation narrows down the targets that the actor may focus on, helping defenders focus their mostly limited defense resources on the most likely cyber threat attack scenarios, as well as shapes intensity and persistence of a cyber threat attack [12]. However, motivation can be different which means ideological, military, financial and many others.
- *Goals*: According to [13] a goal is "a cognitive representation of a desired endpoint that impacts evaluations, emotions, and behaviors". Thus, a goal consists of an overall end state and behavior objectives and plans needed for attaining it, establishing a goal guide behavior (strategy). In present threat intelligence, a goal can be defined as a tuple based on action and object, but work is required to create a consistent taxonomy at an adequate level of detail [14]. It is well known that typical forms of goals are stealing Intellectual Property (IP), damage infrastructure, embarrass competitors, and others.
- *Strategy*: A non-technical high-level description of a planned cyber threat attack. There are typically multiple different ways attackers can achieve their goals, and the strategy defines which approach the attacker should follow. It is assumed that introducing a formal taxonomy, describing relationships between motives, goals, and strategies, would be advantageous for the advancement of threat intelligence, as well as risk assessment processes.
- *Tactics, Techniques, and Procedures* (*TTPs*): Characterize adversary behavior in terms of what they are doing and how they will do it.
- *Tools*: Cyber threat attackers install and use tools within the attacked network. Tools are often modified so that a tool detected and analyzed in a previous security incident might be similar, but not exactly the same in new cyber threat attacks. In this context malware is a sub-category of tools. In addition, tools might be non-malicious software like vulnerability scanners, network scanning tools, and others used for malicious reasons.
- *Indicators of Compromise* (*IoC*): This element represents detective mechanisms describing how to recognize malicious or suspicious behavior. These and other unusual activities enable cybersecurity team staff monitoring a computer system or network to detect malicious intruders early as part of an intrusion detection process. To create an IoC it is desirable to classify indicators a number of ways

like Atomic indicators, Behavioral indicators, and Computed indicators which is often referred to as ABC Indicators [15].

- *Target*: Mostly organizations, companies, sectors, nations, individuals, and others.

And at the preventive level for cyber threat attacks, the following element is of importance:

- *Courses of Action*: Refer to techniques and procedures of the target who try to mitigate what the threat actor wants to achieve. This calls for measures that can be taken to prevent or respond to cyber-attacks.

Mastering the tactics and tools of the Advanced Persistent Threat (APT) hacker, the book in [16] reveals the mindset, skills, and effective attack vectors needed to compromise any target of choice. APT hacking discusses the strategic issues that make all organizations vulnerable and provides noteworthy empirical evidence. In the book [16] one can learn an APT hacker methodology for systematically targeting and infiltrating a public and private organization and its IT systems. A unique, five-phased tactical approach to APT hacking is presented in [16] for real-world examples and hands-on techniques that can be used immediately to execute very effective attacks. By reviewing empirical data from actual attacks conducted by unsophisticated and elite APT hackers, one can learn the APT hacker methodology—a systematic approach designed to ensure success, avoid failures, and minimize the risk of being caught.

3.3 Known-Knowns, Known-Unknowns, Unknown-Unknowns

The analysis of ABC Indicators tries to seek the most static of ABC Indicators but also adversarial behavior because they often reveal themselves. In this regard profiling ABC Indicators typically begin with the most potential assumed cyber-criminal attacker (adversary) as first step and then dovetails into the potential defending strategic options. These options refer to known-knowns (see Table 3.1) with regard to previous cyber threat attack behavioral indicators as cyber threat attack specific signatures out of a crowd of indicators to track. However, profiling in this regard is some kind of a stochastic process and one never knows for sure whether or not the known-knowns that are tracked are identical signatures again behind the actual cyber threat attack. Finally it is only an expectation whether or not the same

Table 3.1 Cyber threat risk level and respective security model

Cyber threat risk level	Security model
Known Knowns (KK)	Information security
Known Unknowns (KU)	Cybersecurity
Unknown Unknowns (UU)	Cyber resilience

adversary (or even adversary group) is truly at the other end of behavior cyber threat attack indicators every time. The adversary group, also called hacker group as a whole, can exist out of numerous individual groups, but all should share the same goals [13]. Based on the received profile it will be possible to facilitate predicting future activity and detecting it in the context of known-unknowns (see Table 3.1).

In Table 3.1 "known knowns" is known cyber threat attack signatures against the CIA Triad (see Chap. 2) based on cyber threat information, which corresponds to methods used in information security. Albeit those security measures demonstrate high detection accuracy for known cyber threat attack patterns, drawbacks include their inability to reliably detect not only unknown but also modified versions of known threats. The "known-unknowns" are Advanced Persistent Threats (APT) or non-CIA Triad cyber threat attacks within the cyberspace, where handling cyber threat attacks requires cooperation between one or multiple stakeholders, who correspond to methods used in cybersecurity (see Chap. 2). The level of "unknown-unknowns" refers to cyber threat attacks which have not yet been identified by anyone and are stated as unforeseeable or unpredictable, which corresponds to cyber resilience. Cyber resilience is a measure of how well a public or private organization can manage a cyber threat attack or data breach while continuing to operate its business effectively. Therefore, the aim of cyber resilience is to ensure that public and private organizations' business operations are safeguarded, and to make sure that cyber threat attacks or data breaches do not demobilize their entire business.

The observation or measure of "known-knowns" cyber threat attack goals to defend and protect data/information assets and communication infrastructures against known cyber threat attacks can be simple as far as they belong to the CIA Triad and Cyber Threat Intelligence (CTI) and will be available. This can be achieved by CTI sharing communities whereby in most cases a well written documentation of how to mitigate the cyber threat attack will be available. In some cases specific configured mitigation tools will allow detecting and even mitigating the cyber threat attack automatically. Google and Amazon for instance have implemented changes in their Domain Name System (DNS) services that inspect Server Name Indication (SNI) fields to detect domain fronting. To detect domain fronting, public and private organizations need the capability to inspect Transport Layer Security (TLS) traffic between internal networks and external hosts.

A way to mitigate cyber threat attacks is to follow up different security alerts such as the ones referred to in Common Vulnerabilities and Exposures (CVE) and/ or National Vulnerability Database (NVD) which often also contains information on how to mitigate the vulnerability. However, achieving a reasonable quality in CTI is a tricky task. It can become a major problem, when public and private organizations and individual security analysts try collaborating to enhance quality by creating, sharing, improving and using CTI as an essential need to improve their cyber defense strategy [6, 17, 18]. Nevertheless private and public organizations need to have their own strategic cybersecurity CTI approach to defend cyber threat attack risk incidents to build up cybersecurity resilience [6, 19]. Such an approach calls for the cybersecurity characteristics as shown in Table 3.1, which deals with the issue on how to handle known and unknown information.

For "known-unknowns" cyber threat identification and observation becomes more difficult and decisions for required CTI with regard to cyber threat attacks will be much harder to achieve. The reason for this lies in the variety of Advanced Persistent Threats (APTs) and the non-CIA Triad cyber threat attacks. The latter requires for example the enhanced and systematic application of the CIA Triad to all assets of the public and private organizations and their environments as an informal description of a cybersecurity solution approach. This requires an End2End (E2E) real-time monitoring for detecting malware by analyzing network traffic to extract network behavioral indicators across different protocols and network layers. This approach refers to different observation methods such as transaction, session, flow and conversation windows. A feature selection method is used to identify the most meaningful features and to reduce the data dimensionality to a tractable size. Finally, various supervised methods are evaluated to indicate whether traffic in the network is malicious, to assign the unknown to the known malware features and to discover new cyber threat attacks [20].

The problem of unknown cyber threat attack detection is a known topic in many different areas in computer systems and network security. Network attacks and malware are among the most common cyber threat attack vectors [21], and are focal points in the dissertation of Duessel [22]. Generally machine learning technique (see Chap. 5) is concerned solving the problem of unknowns by minimizing the expected risk.

Against this background, the category of "unknown-unknowns" refers to the risks of cyber threat attacks which have not yet been identified by anyone. Therefore, they are unforeseeable and unpredictable unknown cyber threat attacks and represent a dynamically changing risk of private and public organizations and their computer systems or networks. This requires a solution for unpredictable implications making the public and private organizations and their computer systems or networks cyber resilient. Unfortunately there are only a few methodological approaches to deal with "unknown-unknowns". One is the so called digital forensic approach also taking into account digital profiling; the other approach is based on machine learning techniques (see Chap. 5). In general terms, digital forensics is the application of scientific investigation techniques to digital crime and cyber threat attacks, but also a crucial aspect of law and business in today's digital world. Therefore, it can be stated as a digital process of preservation, identification, extraction, and documentation of evidence which can be used by the court of law, based on finding evidence from digital systems and devices by forensic teams.

3.4   Digital Forensic and Threat Intelligence Platforms

The scope of digital forensics is to conduct a structured examination through profiling and at the same time to document a chain of evidence so that it is possible to determine exactly which expectations took part through cyber-crime or cyber threat attacked computer systems or networks. This can help answering the question about which adversary (or even an adversary group) potentially may be responsible for the cyber-crime or cyber threat attack. Against this background digital profiling can be understood as a process of gathering and analyzing information about an individual that exists online. Hence, digital profiles include information about personal characteristics, behaviors, affiliations, connections, and interactions. Thus, digital profiling can be used in a manifold of areas. In public and private organizations, security through digital profiling is used to identify suspect employees and protect the public and private organization from an inside cyber-crime attack. To determine whether the identified employee really poses a risk to the public and private organization, his online behavior may be scrutinized as a digital profile. In a low-profile incident case, information would typically be gathered through corporate email, logs and social media content, connections and posts. In a more high-profile incident case, investigators might employ specific surveillance technologies for a more complex profile of the individual.

Forensic approaches use various techniques and proprietary forensics applications to examine the infiltrated computer system or network by searching hidden folders and unallocated space on volumes of deleted, encrypted or corrupted files. Evidence to the unknown that is found will be carefully documented in a Forensic Investigation Profile Document (FIPD). Determining a forensic profile for unknown cyber threats contains several steps such as:

- Identifying and documenting the highest probability of a potential nature and purpose for an unknown cyber-crime or cyber threat attack,
- Identifying and documenting the highest probability of a potential unknown cyber-crime or cyber threat infection mechanism,
- Identifying and documenting the highest probability of how an unknown cyber-crime or cyber threat attack may interact with which potentially targeted host computer system or network,
- Identifying and documenting the highest probability of the profile and sophistication level of an unknown cyber-crime or cyber threat attack,
- Identifying and documenting the highest probability of the extent of infection and compromise of the potentially targeted host computer system or network by an unknown cyber-crime or cyber threat attack,
- And many others.

Based on this procedure specific digital forensic incident response programs can be built up, which combines the respective cybersecurity tools and approaches with regard to the different forms of cyber-crime and cyber threat attacks to respond effectively. The portfolio of tools and approaches in digital forensics require specific skills such as

- Reverse-engineering of malware
- Detecting malicious files and software code
- Discovering, searching computer systems memory for infections and malicious code
- Digital documents for infections and cyber threats

- And many others

These tools and approaches come in action both before and after a breach. The tools and approaches portfolio includes endpoint detection and response monitoring, cybersecurity information and incident management, log analyzers, threat intelligence databases, penetration testing, firewalls, intrusion detection and prevention (see Chap. 4), machine learning (see Chap. 5), and many others.

Finally the sums off all evidences form the puzzle to be analyzed to identify how to transform the unknown cyber threat attack to an understandable new cyber threat attack form. For this reason, there are several efforts available to find a uniform method in the documentation and incident reports, such as:

- *Open Indicators of Compromise (OpenIoC)*: OpenIOC is an open framework designed to fill a void that currently exists for organizations that want to share cyber threat information both internally and externally in a machine-digestible format, developed by the American cybersecurity company MANDIANT in 2011 [23]. It is written in eXtensible Markup Language (XML) and can easily be customized for additional intelligence that incident responders can translate their knowledge into a standard format. This allows private and public organizations to describe technical characteristics that identify a known cyber threat, an attacker's methodology, or other evidence of compromise. Thus, organizations can leverage this format to share cyber threat-related latest Indicators of Compromise (IoCs) with other organizations, enabling real-time protection against the latest cyber threats [24].

- *Structured Threat Information eXpression (STIX™)*: Standardized structured language for describing Cyber Threat Intelligence (CTI) developed by MITRE Corporation and the Open Standards Open Source (OASIS) Cyber Threat Intelligence Technical Committee which supports automated information sharing for cybersecurity situational awareness, real time network defense, and sophisticated cyber threat analysis. In STIX™ terminology an individual or group involved in malicious cyber-crime activity is called a *Threat Actor*. A set of activity (*Incidents*) carried out by *Threat Actors* using specific techniques (*TTP*) for some particular purpose is called a *Campaign*. Such activity might fit along the lines of stealing data from customers or targeting a particular business sector. When data is collected on various related intrusion attempts (*Incidents*), it may not initially include enough data for characterizing attribution of the actor causing them. In this case, for cross-incident analysis of the "who" and "why", the preferred method is to begin by defining a *Campaign* for that activity with a placeholder *Threat Actor* identity until additional information comes to light. As more information evolves for characterizing the responsible actors the *Threat Actor* placeholder can be incrementally fleshed out [25]. The STIX™ relationship is illustrated in Fig. 3.1 after [10], whereby the IoC contains a pattern that can be used to detect suspicious or malicious cyber threat activity, cyber threat attacker refers to individuals, groups, or organizational attackers believed to be operating with malicious intent, campaign groups adversarial behaviors that describes a set of malicious activities or cyber threat attacks that occur over a

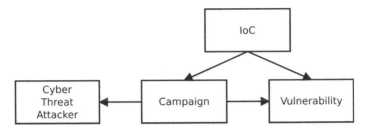

Fig. 3.1 STIX™ relationship example

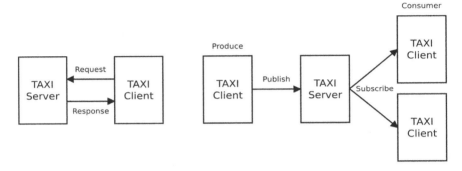

Fig. 3.2 TAXII™ collections (**a**) and channels (**b**)

period of time against a specific set of targets, and vulnerability is assumed as a mistake in software that can be directly used by a hacker to gain access to a system or network. STIX™ is designed to allow users to describe threat attacks [10, 26, 27].

- *Trusted Automated eXchange of Indicator Information* (*TAXII*™): Application layer protocol for the communication of cyber threat information in a simple and scalable manner, used to exchange Cyber Threat Intelligence (CTI) over HTTPS. TAXII™ enables public and private organizations to share CTI by defining an Application Programming Interface (API), a set of services and messages as requirements for clients and servers, supporting the common sharing models collections and channels, as illustrated in Fig. 3.2 after [10] whereby a collection is an interface to a logical repository of CTI objects provided by a TAXII™ Server that allows an operation team to host a set of CTI data that can be requested by users or groups: TAXII™ clients and servers exchange information in a request-response model, and a channel, maintained by a TAXII™ server, allows operation teams to push data to many users and user groups to receive data from many threat operation teams: TAXII™ clients exchange information with other TAXII™ clients in a publish-subscribe model. Note: The TAXII 2.0 specification reserves the keywords required for channels but does not specify channel services. Channels and their services will be defined in a later version of TAXII™.

TAXII™ is a free and open framework that standardizes the automated exchange of cyber threat attack information through services and messaging, designed to integrate with existing sharing agreements, including access control limitations. TAXII™ has been transitioned to OASIS which supports automated information sharing for cybersecurity situational awareness, real-time network defense, and sophisticated cyber threat analysis. It has been specially developed to support STIX™ information; this is done by defining an Application Programming Interface (API) based on common exchange models [26].

Several companies that have business in cybersecurity like Anomali, EclecticIQ, Fujitsu, Hitachi, IBM Security, New Context, NC4, ThreatQuotient, and TruSTAR are demonstrating how STIX and TAXII™ are being used to prevent and defend against intruding cyber threat attacks by enabling threat intelligence to be analyzed and shared among trusted partners and communities [28].

3.5 Threat Attack Profiling, Threat Intelligence, and Threat Lifecycle

Cyber threat attacks can be described by a sequence of phases, as shown in Fig. 3.3. From Fig. 3.3 it can be seen that if the phases shown are completed successfully, compromising code is present on a computer system or network that should not be there, and intrusion of a cyber threat attack is not defended to avoid negative impacts, for example, exfiltration of data/information which has been the target all along.

The Command-and-Control phase (C2) of the attack represents the period after which the cyber-attacker leverages the exploit of a computer system or network. This shows the importance to analyze cyber threat attacks taking into account the different indicators to the phases illustrated in Fig. 3.3. Assuming that a cyber threat attacker may attempt to use the zero-day attack to compromise targeted computer systems or networks, C2 may be the same as past cyber threat attacks by the same cyber-attacker. Different proxy IP addresses can be used to relay a cyber threat attack, but the intrusion detection tools used may not change between these cyber threat attacks. These immutable or infrequently-changing properties of cyber threat attacks by a cyber threat attacker form a cyber threat attacker's behavioral profile. This profile refers to the important knowledge that capturing, knowing, and detecting the specific intrusion form that facilitates discovering of other cyber threat attacks by the same cyber threat attacker, even if many other aspects of the cyber

Fig. 3.3 Cyber threat attack progression sequence

threat attack change. However, cyber threat attack analysis quickly becomes complicated, and will make profiling cyber threat attacker much more complex, because analysis of attack progression requires knowing the cyber threat attack Indicator of Compromise Lifecycle (IoCL). However, the IoCL has an intrinsic cycle which means that with the discovery of known cyber threat attacks IoC the revelation of new ones will start. Therefore, knowledge is of importance to decide if a detected cyber threat attack IoCL will further be active or not and has therefore to be taken into consideration. In this regard IoCL refers to answering the intrinsic questions of how the collecting, analyzing and disseminating data sets or data packets of cyber threat attacks and vulnerabilities incidents work.

Against this background Cyber Threat Intelligence (CIT) emerged in order to support cybersecurity practitioners in recognizing the indicators of cyber threat attacks, extraction information about cyber threat attack method, and consequently responding to cyber threat attack accurately and in a timely manner. However, a salient problem in CTI is shortening time for improving decision making for an increasing volume of possible cyber threat alerts, which has to be fast. Alerts provide timely information about current security issues, vulnerabilities, and exploits. Many alerts can be classified as unknown so far, not investigated and mark a type of uncertainties, which can be captured by their ontologies (see Chap. 7) or events that have not been thought of. To reduce response times, cyber threat attack incident responses must become less reactive and more proactive. This requires additional methods, for instance filter out false alarms, speed up triage (see Table 3.2), and simplify cyber threat attack incident analysis. If suspicious incidents are detected through decision making, CTI algorithms must be executed immediately to reduce the impact and severity of the cyber threat attack. These algorithms can be based on machine learning (see Chap. 5), static malware analysis, classification algorithms, data sets and data packets analysis, as well as feature selection algorithms, and others [29]. Thus, CTI algorithms deal with cybersecurity operations, as illustrated in Table 3.2.

Reducing response time to cyber threat attack incidents, responses must be proactive, detecting probable cyber threat attacks and periodization, strengthening cyber threat attack incident response through threat intelligence detection and cyber threat defense algorithms. The objective of these algorithms is to accurately discover suspicious incidents based on the analysis of actual measured data to defend cyber threat attack vulnerability in real time. However the objective is hard to achieve, especially in terms of accuracy.

Table 3.2 Stages versus tasks

Stage	Tasks
Triage	Determine importance and integrity of reaction to incoming alerts
	Decision if alert is legitimate and should be tracked
First response	Determine scope of incident identified
	Identify infected and vulnerable components, computer systems or networks
	Recommend actions to contain the effects
Investigation	Determine strategic weaknesses in defending threat incidents
	Recommend action to prevent recurrences

If threat intelligence detection and defense algorithms generate inaccurate results, the outcome can negatively impact the performance of the entire Threat Intelligence Defense System (TIDS) due to unmanageable numbers of alarm notifications that may overwhelm security nodes. Consequently, research has explored new algorithms and methodologies aiming to increase the performance and accuracy of cyber threat intrusion detection systems. However, there is a need for effective intelligence management platforms to facilitate the generation, refinement, and vetting of data, post sharing. Designing such a system, some of the key challenges are working with multiple intelligence sources, combining and enriching data for deeper knowledge, determining intelligence relevance based on technical constructs, and organizational input, delivery into organizational workflows and finally into technological products.

3.6 Threat Intelligence Sharing and Management Platforms

A threat intelligence platform supports cyber threat defending teams, security operations centers, as well as threat intelligence analysts, responsible for incident response, risk management and vulnerabilities, so that they not only have to react to events and warnings, but can also predict cyber threat attacks and act more proactively. To achieve this, the threat intelligence platform serves as a central source of information for all cyber threat data from external and internal sources. This simplifies collaboration, informed decisions, proactive measures, and faster detection and response.

Brown et al. [30] discuss the community requirements and expectations of an all-encompassing threat intelligence management platform (TIMP) based on studies on a few threat intelligence sharing platforms. As the threat landscape and the public and private organizations environments will change, the threat intelligence platform also helps to stay up to date with potential cyber threats, predict them, and adapt to ongoing cyber threat assessments, and strengthening countermeasures. However, finally the question to be answered which is still paramount, is whether a threat intelligence management platform is to be bought or built, to support, identify, prioritize, and act on the most relevant cyber threats to the public and private organizations' business to secure them against cyber threat attacks. In this context some important considerations have to be taken into account deciding building or buying. On the one hand it is a question of human resources because it can be difficult for public and private organizations to find the required staff with the essential knowledge necessary to build up such a central software system platform and keep this resource in the long-term continuously up-to-date. On the other hand this kind of software development is more than only a software. The software developer(s) must be familiar with software architectures, development processes (DevOps), security processes (SecOps) and security technologies, friendly speaking a combination which is extremely rare to find. This means that the essential employees have to be at hand. To build such a development team to support public and private business operations requires several full-time employees:

software developer(s), database/big data expert(s), cybersecurity expert(s), and quality assurance expert(s) being able to aggregate and carry out internal analyses of malware data, vulnerability data and risk management factors is essential. Against this background, it could be much easier for public and private organizations to buy a threat intelligence management platform (TIMP) framework to hire new employees.

References

1. A. Malatras, C. Skouloudi, A. Koukounas, *Industry 4.0—Cybersecurity Challenges and Recommendations* (European Union Agency for Cybersecurity (ENISA), Heraklion, 2019)
2. M. Lehto, Phenomena in the cyber world, in *Cyber Security: Analytics, Technology and Automation*, ed. by M. Lehto, P. Neittaanmaki, (Springer Publ., Cham, 2015), pp. 3–30
3. T. De Zan, F. Di Franco, *Cybersecurity Skills Development in the EU* (European Union Agency for Cybersecurity (ENISA), Heraklion, 2019)
4. S. Barnum, *Standardizing Cyber Threat Intelligence Information with the Structured Threat Information eXpression (STIX™)* (MITRE Org., Bedford, MA, 2014)
5. R. McMillan, *Definition: Threat Intelligence* (Gartner Research, Stamford, CT, 2013)
6. J. Robertson, A. Diab, E. Martin, E. Nunes, V. Paliath, J. Shakarian, P. Skakarian, *Darkweb Cyber Threat Intelligence Mining* (Cambridge University Press, Cambridge, 2017)
7. R. Borum, J. Felker, S. Kern, K. Demnesen, T. Feyes, Strategic cyber intelligence. Inform. Comput. Security **23**(3), 317–332 (2015)
8. https://nvd.nist.gov/
9. https://cve.mitre.org/
10. https://oasis-open.github.io/cti-documentation/
11. V. Mavroeidis, S. Bromander, Cyber threat intelligence model: an evaluation of taxonomies, sharing standards, and ontologies within cyber threat intelligence, in *Proceedings of the European Intelligence and Security Informatics Conference*, 2017, pp. 91–98
12. T. Casey, Understanding cyber threat motivations to improve defence, Intel White Paper, 2015
13. A. Fishbach, M.J. Ferguson, The goal construct in social psychology, in *Social Psychology: Handbook of Basic Principles*, ed. by A.W. Kruglanski, E.T. Higgins, (The Guilford Press, New York, 2007), pp. 490–515
14. S. Bromander, A. Josang, M. Eian, Semantic cyberthreat modeling, in: *STIGDS*, 2016, pp. 74–78
15. SANS, Security Intelligence: Attacking the Cyber Kill Chain. https://digital-forensics.sans.org/blog/2009/10/14/security.intelligence-attacking-the-kill-chain/
16. T. Wrightson, *Advanced Persistent Threat Hacking: The Art and Science of Hacking any Organization* (McGraw-Hill, New York, 2010)
17. O. Al-Ibrahim, A. Mohaisen, C. Kamhoua, K. Kwiat, L. Njilla, Beyond Free Riding: Quality of Indicators for Accessing Participation in Information Sharing for Threat Intelligence, Technical Report University at Buffalo and Air Force Research Lab, 2017. https://arxiv.org/abs/1702.00552
18. C. Sillaber, C. Sauerwein, A. Mussmann, R. Breu, Data quality challenges and future research directions in threat intelligence sharing practice, in *Proceedings of the ACM Workshop on Information Sharing and Collaborative Security*, 2016, pp. 65–70
19. G. Sharkov, From cybersecurity to collaborative resiliency, in *Proceedings of the ACM Workshop on Automated Decision Making for Active Cyber Defense*, 2016, pp. 3–9
20. D. Bekerman, B. Shapira, L. Rkach, A. Bar, Unknown malware detection using network traffic classification, in *Proceedings of the IEEE Conference on Communications and Network Security (CNS)*, 2015, pp. 134–142

21. P. Fogla, M. Sharif, R. Perdisci, O. Kolessnikov, W. Lee, Polymorphic blending attacks, in *Proceedings of the 15th USENIX Security Symposium*, 2006, pp. 241–256
22. P. Duessel, Detection of Unknown Cyber Attacks Using Convolution Kernels over Attributed Language Models, PhD Thesis, University of Bonn, Bonn, 2018
23. https://cyware.com/educational-guides/cyber-threat-intelligence/what-is-open-indicators-of-compromise-openioc-framework-ed9d
24. H.-Y. Lock, *Using IOC (Indicators of Compromise) in Malware Forensic* (SANS Institute, Bethesda, MD, 2019)
25. https://stixproject.github.io/documentation/idioms/campaign-v-actors/
26. https://www.anomali.com/de/what-are-stix-taxii
27. R. Struse, J. Wunder, M. Davidson, B. Jordan, *TAXI™ Version 2.0 Working Draft 02* (OASIS Open, Burlington, MA, 2017)
28. https://www.oasis-open.org/news/pr/cybersecurity-companies-demo-support-for-stix-and-taxii-standards-for-automated-threat-intel?platform=hootsuite
29. A. Dehghantanha, M. Dargahi (eds.), *Cyber Threat Intelligence* (Springer Publ., Cham, 2015)
30. S. Brown, J. Gommers, O. Serrano, From cybersecurity information sharing to threat management, in *Proceedings of the 2nd ACM Workshop on Information Sharing and Collaborative Security*, 2015, pp. 343–349

Chapter 4
Intrusion Detection and Prevention

4.1 Intrusion Detection

Intrusion detection is a methodological approach establishing cybersecurity in existing computer systems and networks because they are often operated in an open task mode. Therefore, the goal of intrusion detection is to detect, preferably in real time, unauthorized access or use, misuse, and abuse of computer systems or networks, by both, computer system or network insiders like enterprise employees, as well as external penetrators with their cyber threat attacks. Thus, intrusion detection is becoming a challenging task due to the proliferation of heterogeneous computer systems and networks with the increased connectivity of computer systems and networks in public and private organizations, because connectivity allows easy accessibility to outsiders. Hence, the main task of an Intrusion Detection System (IDS) is to defend computer systems and networks by detecting hostile cyber threat attacks or exploits in a computer system or network. This requires monitoring the data flows or packets occurring in a computer system or network and analyzing them for signs of suspicious activity and finally report possible security incidents. While anomaly detection and reporting is the primary function of an IDS the intrusion detection software system automates the Intrusion Detection Process (IDP) taking the respective actions when malicious activity or anomalous traffic is detected, including blocking data flow or packet traffic sent from suspicious IP addresses. Typically the intrusion detection system is placed inline, at a spanning port of a switch, or on a hub in place of a switch.

Following [1] intrusion detection can be introduced as any set of actions to identify attempts to compromise the CIA Triad (see Sect. 2.1) of private and public organizations' computer systems and network resources. An intrusion detection system passively monitors for attacks and provides notification services for active intrusion defense requirement in case of identified anomalies. In this context a cybersecurity breach can be introduced as a violation of the cybersecurity constraints of computer system and network resources with the scope to corrupt the actual executed code, for instance by intruding malicious code. In this regard Intrusion Detection Systems (IDSs) goal is to detect any set of cyber threat attack intrusion incidents in a computer system or network, and sending out an alert which inform whenever the computer systems or networks cybersecurity has been compromised by a cyber threat attack. However, depending on the placement and the methodology deployed for cybersecurity in a computer system or network, different kinds of IDS can be distinguished. The mostly used approaches are Host-based Intrusion Detection System (HIDS), Network-based Intrusion Detection System (NIDS), and Specification-based Intrusion Detection System (SIDS) [2–4]. After detecting an intrusion incident the generated alert is transferred to an entity that responds to the alert and takes the appropriate action to start with defending the intrusion by ousting the intruder. But the problem with intrusion detection is that a cyber threat attack incident can be one out of a number of different types. For example, in one case, an unauthorized user might steal passwords to masquerade his true identity to the attacked computer system or network. Another type of cyber threat attack intruders are people who are legitimate users of computer systems or networks, but who abuse their privileges, as well as people who use prepacked exploit scripts; which often can be found on the Internet, to attack computer systems through a network. Therefore, Intrusion Detection and Intrusion Prevention (IDP) is required that consists of a data collection device that collects data or packet traffic to be monitored to enhance detection quality. However, any definition of an intrusion detection type is imprecise as security policy requirements do not always translate into a well-defined set of actions. Hence, intrusion detection can be stated as a general methodology by which intrusions can be detected. An Intrusion Detection System (IDS) can be located inline of the legitimate data or packet traffic to public or private organizations' computer systems or network environments, to monitor all internal legitimate data or packet traffic flow. The models for positioning the IDS module in legitimate network traffic is described in [3] consider using both, a Firewall and IDS.

Since, an IDS generates a large amount of data, traffic flow and events in its logs, the main feature for IDS is to generate alerts on incidents of interest and danger. Thus, effective IDS should have a low rate of false positives and false negatives, as described in [5], and shown in Table 4.1.

For IDS different methodological approaches and classifications exist. This classification contains HIDS, NIDS, and SIDS. Besides these approaches two other intrusion detection approaches are used [4–8] and are broadly classified as:

Table 4.1 True and false positive and negative

	Positive	Negative
True	Alert when there is malicious data or traffic flow	Silent when data or traffic flow is rare or unlikely (see Table 2.2)
False	Alert when data or traffic flow is rare or unlikely (see Table 2.2)	Silent when malicious data or traffic flow occurs

- *Anomaly Intrusion Detection (AID)*: Protect computer systems or networks against malicious incidents.
- *Misuse Intrusion Detection (MID) or Signature Intrusion Detection (SID)*: Protect computer systems or networks based on patterns against suspicious collections of sequences of activities or operation that possibly are harmful.

4.1.1 Anomaly Intrusion Detection

A behavior that is neither nominal nor normal is described as anomalous. Hence, anomaly detection is a key to identify items, incidents, or observations that do not confirm to an expected pattern or other items in a dataset or packet. Examples of anomalies are for instance data points not following a particular distribution, the occurrence of incorrect or infrequent values, or the repeated presence or absence of particular events, and many others. Thus anomaly intrusion detection classifies deviations from normal behavior that indicate the presence of intentionally or unintentionally accomplished suspicious attacks or cyber-criminal fraud. Thus, anomaly detection focuses on detecting activity patterns in the observed data. Hence, an Anomaly Intrusion Detection System (AIDS) must be able to distinguish between normal and anomalous behavior. Hence, anomaly detection approaches are based on models of normal data sets or packets that detect deviations and monitor normal data sets or packets. Beside this, anomaly detection must also be capable of detecting new types of cyber threat attack-based intrusion incidents such as deviations from normal data transmission usage. In this regard it is suitable to divide anomaly detection systems into static and dynamic anomaly detection systems.

Furthermore, anomaly detection can be either time-dependent or time-independent. Time-dependent anomaly detectors focus on the detection of anomalies in temporal data sets by taking time data into account for instance time between events, time of occurrence, or event ordering. Time-independent anomaly detectors ignore temporal information and focus on detection of anomalies on individual data points, for instance, inside multivariate data sets of an event, or data aggregations [7].

Anomaly Intrusion Detection System (AIDS) often uses a Self-Organizing Map (SOM) algorithm to model signatures of normal data or packet traffic to determine if a computer system or network connection data flow is normal or abnormal. Alerting and reporting handle communication based on a decision support system. The SOM algorithm belongs to the category of competitive learning models, commonly successfully used for various clustering problems [9]. The SOM method can be based on unsupervised learning to map nonlinear statistical relationships between high-dimensional input data into two-dimensional space, the so called output space. SOMs efficiently place similar signatures to adjacent locations in the output space and provide projection and visualization options for high dimensional data. To train the SOM algorithm, the first step is to enumerate and normalize input vectors which can be accomplished in a pre-processing submodule. The input data sets for training contain normal data sets and data obtained from cyber threat attacks or abnormal signature data or packet flows. The anomaly detection submodule extracts pre-processed data sets out of the normal data sets that are used to train the SOM. After a successful training phase the SOM is ready to be used for classification of normal or abnormal behavior.

In this regard with AIDS, activities are monitored which periodically generate signatures capturing their behavior. As monitored input data sets or packets are processed, the AIDS periodically generates a value to indicate normal or abnormal behavior. In case of an anomaly where too much deviation from the normal signatures occurs, the Intrusion Detection Decision Support System (IDDSS) reports an alert due to an identified cyber threat attack intrusion incident. However, this can lead to false positive alarms meaning that a monitored regular functional operation is mistakenly identified as a cyber threat attack due to an error, and that the AIDS, as a result of the fault, stop the regular functional operation. This is likely a negative impact on computer system or network function, depending on the conditioning or sensitivity of the AIDS. Hence, false positives are events that are reported as malicious but in reality they are not. Therefore, logging is an important issue of anomaly intrusion detection to record intrusion-related activities.

However, a lack of anomaly intrusion detection allows a cyber-attacker to attempt cyber threat attacks until a successful one is identified. Anomaly intrusion detection allows the cyber threat attack to be identified long before a cyber threat attack is likely successful. A simple rule-of-thumb is that: If the computer system or network could not have reasonably been generated by a legitimate user of the application, it is almost certainly a cyber-attack incident. Once alerted by the AIDS and the cyber threat attack is identified, the Intrusion Prevention Systems (IPS), which is next in line, can respond appropriately to the alert. This typically includes logging off the user from where the anomalous data traffic is generated or intruded, invalidating the account, recording information for the cybersecurity team or patching the root cause vulnerability. To find an appropriate solution, the safest rule is to assume that everything except legitimate traffic is a cyber threat attack. However, this will probably cause false alerts which can be false positive as well as false negative, and hence hard to deal with. As mentioned earlier in case of false positive the AIDS identifies an incident as a cyber threat attack, albeit the incident is an acceptable behavior. Nevertheless, a false positive is a false alert. A false negative incident is the most serious and dangerous alert that could happen, when the AIDS identifies an incident as acceptable and this incident actually is a cyber threat attack. Therefore, a false negative alert is an incident, where the AIDS fails to detect a real cyber-attack incident. This is the most dangerous cyber threat alert because it informs that a computer system or network behavior, albeit a cyber threat attack took place. Against this background, anomaly detection includes the following advantages and disadvantages:

- *Advantage of Anomaly Intrusion Detection*: No predefined rules for detection of intrusions are required; hence new cyber threat attacks can be detected.

- *Disadvantages of Anomaly Intrusion Detection*: False positive alert can arise, leading to inconvenience for the users. Establishment of regular profile usage is required but is often hard to achieve.

Furthermore, anomaly detection includes the following types of models [10]:

- *Statistical Models (SM)*: These techniques make use of different kinds of models such as operational model or threshold metric, Markov model or marker model, statistical moment or mean and standard deviation model, multivariate model, time series model.
- *Cognitive Models (CM)*: These techniques make use of different kinds of methods such as finite state machine method, description scripts describing signatures of attacks, adept process management model which is trained by a huge number of patterns with known attack patterns.
- *Cognitive based Detection Techniques (CDT)*: This technique makes use of audit data classification technique based of sets of predefined rules, classes and attributes identified from training data sets of classification rules, parameters and procedures inferred and others as illustrated in Fig. 4.1 after [10].

4.1.2 Misuse Intrusion Detection

A multitude of technical safeguards is available to detect and prevent cyber threat attacks. Concepts behind are broadly categorized into the anomaly detection described in Sect. 4.1.1, misuse detection as described in this section, and specification based intrusion detection described in Sect. 4.1.4. Misuse detection methods are intended to recognize known attack signatures described by rules. Thus misuse detection methods are intended to recognize known attack patterns described by rules. The majority of the state-of-the art methods can be classified as misused detection due to their reliance on rule sets. Rule-based solutions can be divided into

- *Blacklist-based Method (BM)*: Blacklisting has been deployed as a key element in anti-virus and security software suites, typically in the form of a so called virus database of known digital signatures, heuristics or behavior characteristics associated with viruses and malware that have been identified in the wild. It can be differentiated into the signature-based approach and the heuristic-based approach.
- *Whitelist-based Method (WM)*: Whitelisting draws up a list of acceptable entities that are allowed access to a computer system or network, and blocks everything else. It usually includes policies which allow cyber threat attacks detection based on pre-defined negative baseline configuration, IP whitelists.

The majority of intrusion detection systems can be classified as misuse detection. In this regard misuse detection tries to match actual data and traffic flow activity to stored signatures of known exploits or cyber threat attacks which means that Misuse Intrusion Detection (MIS) uses a prior knowledge on cyber threat attacks to

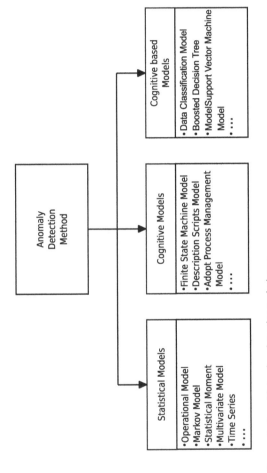

Fig. 4.1 Classification of anomaly based intrusion detection techniques

investigate for cyber threat attack traces. This is based on well-defined signatures of input incidents, assuming that the state transition of the computer system or network leads to an intruded state when exercised with the intrusion pattern, weaknesses in the computer system or network and application software, which can be exploited. The objective of MIS is to frame the intrusion detection problem as a signature or pattern-matching problem and to develop efficient algorithms for such a matching problem. But simply specifying an intrusion pattern without the initial state specification is often insufficient to capture an intrusion scenario. Other primary approaches to misused detection techniques are [6]:

- *Expert Systems (ES)*: Code knowledge about cyber threat attacks is based as if-then implication rules,
- *Model-based Reasoning System (MRS)*: Combine models of misuse with evidential reasoning to support conclusions about the occurrence of misuse,
- *State Transition Analysis (STA)*: Represents cyber threat attacks as a sequence of state transitions of the monitored computer system or network and
- *Key Stroke Monitoring (KSM)*: Uses user key strokes to determine the occurrence of a cyber threat attack.

Another misuse detection method makes use of cyber threat attack signatures, described by rules. These rule based methods can be divided into blacklist- and whitelist-based approaches. Blacklist-based methods being further refined into signature-based and heuristic-based approaches:

- *Signature-based Misuse IDS*: Based on defined signatures in order to detect known cyber threat attacks; for instance, detecting threats based on specific cyber threat attack signatures for instance malicious byte sequences. As a result signature based IDS are effectively detecting cyber threat attacks without too many false alarms. However, signature based IDS being unable to detect cyber threat attacks whose signatures are unknown.
- *Heuristic-based IDS*: Method that allows detection of unknown cyber threat attacks based on an expert-based probabilistic rule sets that describe malicious indicators. However heuristic approaches often complement signature-based solutions with regard to their susceptibility to high false positive rates.

In contrast white-list based approaches usually include policies which allow the detection of cyber threat attacks based on the deviation from a pre-defined negative baseline configuration, for instance IP whitelists.

To overcome challenges with existing approaches, for instance limited ability to detect unknown cyber threat attacks by signature-based methods and lack of detection accuracy by behavior-based methods, interest has grown in the security community to utilize machine learning as an alternative approach (see Chap. 5).

4.1.3 Disadvantages of Anomaly and Misuse Intrusion Detection

A primary disadvantage of anomaly detection is that statistical measures of user behavior can be gradually trained. Cyber threat attackers who know they are being monitored can train such computer systems or networks over a length of time to the point where intrusive behavior is considered normal. Misuse detection, that is simpler than anomaly detection, is immune to such training. If signatures for a cyber threat attack are carefully written even major variations of the same cyber threat attack scenario can be detected, but this is simpler in misuse detection compared to anomaly detection.

Another disadvantage is that these methods look only for known vulnerabilities, and are currently of little use in detecting unknown future intrusions. However, modern approaches enhance the capability in anomaly intrusion detection by using a SOM structure to model normal behavior, whereby deviation from normal behavior is classified as a security violation, and hence a cyber threat attack [9]. A SOM efficiently places similar patterns to adjacent locations in the output space and provides projection and visualization options for high dimensional data sets [11]. In the two-dimensional (2D) case the artificial neurons can be arranged either on a rectangular or a hexagonal lattice. Each artificial neuron of the SOM has an associated n-dimensional prototype vector

$$m_i = \left[m_{i1}, m_{i2}, \ldots, m_{in} \right]$$

where n is equal to the dimension of the input vectors. Artificial neurons that are adjacent belong to 1-neighborhood N_{i1} of the artificial neuron i. A neighborhood function determines how strongly artificial neurons are connected to each other. In this regard the neighborhood functions and the number of artificial neurons determines the accuracy and the generalization capability of the SOM mapping.

To overcome challenges with existing approaches, for example limited ability to detect unknown cyber threat attacks by signature-based methods and lack of detection accuracy by behavior-based methods, interest has grown in the security community to utilize machine learning (see Chap. 5) as an alternative approach. An overview of state-of-the art anomaly-based intrusion detection and misuse-based intrusion detection methods is provided in the works [12, 13].

4.1.4 Specification-Based Intrusion Detection

Specification-based Intrusion Detection Techniques (SIDT) has been proposed as a promising alternative that combines the strengths of anomaly and misuse detection. In this approach, manually developed specifications are used to characterize legitimate program behavior. As this method is based on legitimate behavior, it does not generate false alarms when unusual, but legitimate program behavior is encountered. Thus, its false positive rate can be comparable to that of misuse detection. Since it detects attacks as deviations from legitimate behaviors, it has the potential to detect previously unknown attacks.

4.1.5   Intrusion Type Characteristics and Detection

In order to classify breaches by cyber threat attacks, a scheme based on intrusion types is presented in [14, 15] and summarized in Table 4.2, which introduces some important intrusion types, their characteristics, and detection possibilities.

Table 4.2 Intrusion types and their detection

Intrusion type	Characteristics	Detection
Attempted break-in	Breaking into systems generates a high abnormality rate of password failures with regard to a single account or the system as a whole	Atypical behavior profile of violation of security constraints
Denial of service	Intruder is able to monopolize a resource and might have abnormally high activity with the resource, while activity for all other users is abnormally low	Atypical behavior profile or violation of security constraints
Inference by legitimated user	User attempting to take unauthorized data from a database through aggregation and inference might retrieve more records than usual	Atypical behavior profile using I/O resources
Leakage by legitimated user	User trying to leak sensitive documents, might lock into system at unusual time or route data to remote printers not normally used	Atypical behavior using of I/O resources
Masquerading by successful break-in	Login through unauthorized account or password, might have different login time, location, or connection type from account's legitimate user. Intruder's behavior may differ from that of a legitimate user e.g. a user using most of his time browsing through directories and executing system status commands whereas legitimate user might edit, compile, or link programs	Atypical behavior profiles or violation of security constraints
Trojan horse	A program is submitted for legitimate program	Atypical using of I/O resources or I/O activity
Virus	May cause an increase in frequency of executable files rewritten or storage used by execution files	Atypical using of I/O resources or I/O activity

Let A_1, A_2, ..., A_n be n measures used to determine if an intrusion by a cyber threat attack is occurring to a computer system or network at any given moment, whereby each A_i measures a different aspect of the computer system and/or network with

$$A_i = \begin{cases} 1 & \textit{implying that the measure is anomalus} \\ 0 & \textit{otherwise} \end{cases}$$

Let H be the hypothesis that the computer system or network is currently undergoing an intrusion by a cyber threat attack. The reliability and sensitivity of each anomaly measure A_i is determined by

$$p\left(A_i = 1 \middle| H\right)$$

and

$$p\left(A_i = 1 \middle| /H\right).$$

The combined belief in H is

$$p\left(H \middle| A_1, A_2, \ldots, A_n\right) = p\left(A_1, A_2, \ldots, A_n \middle| H\right) \times \frac{p(H)}{p\left(A_1, A_2, \ldots, A_n\right)}$$

which requires the joint probability distribution of the set of measures conditioned on H and $/H$. In [16], covariance matrices are used to account for the interrelationships between measures. If the measures A_1, A_2, ..., A_n are represented by vector A, then the compound anomaly measure is determined by

$$A^T C^{-1} A$$

where C is the covariance matrix representing the dependence between each pair of anomaly measures A_i and A_j. The foregoing methodology on intrusion detection is now broadened by the issue of intrusion prevention, the process of performing intrusion detection and defending possible detected threat incidents. Therefore, the issue is introducing intrusion detection and prevention systems that are primarily focusing on identifying possible cyber threat attack incidents, logging information about them, attempting to defend them, reporting them to responsible security administrators in public and private organizations, and documenting existing cyber threats. Hence, intrusion detection and prevention have become an essential issue to the security infrastructure of nearly every mission-critical and crucial computer system or network. The types of Intrusion Detection and Prevention System (IDPS) techniques can be differentiated by the types of potential incidents that they monitor and the ways in which they are deployed, as shown in Table 4.3.

Table 4.3 Intrusion detection and prevention system types

IDPS type	Characteristics
Host-based	Monitoring characteristic of single host and events occurring with that host for suspicious activity
Network-based	Monitoring network traffic for particular network segments or devices, analyzing network and application protocol activity to identify suspicious activity
Network behavior analysis	Examining network traffic, identifying threats that generate unusual traffic flows like Distributed Denial of Service (DDoS) attacks, certain forms of malware, policy violation, e.g. client system providing network services to other networks
Wireless	Monitoring network traffic, identifying suspicious activity involving wireless network protocols themselves

4.1.6 Intrusion Detection System Architecture

Securing critical and crucial computer systems or networks is a very important issue because cyber-attackers want to gain access to sensitive data or information of critical and crucial computer systems or networks and their configurations, vulnerabilities, and others. Therefore, specific protective actions are of particular importance, such as encryption and other actions, for transmitting data or information physically or logically over separate computer system or network components. This includes verifying that the components are working as desired and not anomalous which requires monitoring for security issues, performing regular vulnerability assessments, responding appropriately to vulnerabilities, and testing and deploying of an Intrusion Detection System (IDS). In such IDS data sets are stored and processed directly, and the output of which is fed into a rule based intrusion detector which in turn takes further action. Hence, monitoring is required to identify causes of cyber threat attacks that require an alert. In this regard an alert list of eligible cyber threat attacks is created, and the computer system or network the cyber-attacker is attempting to intrude is locked through intrusion notification alerting, which also report the alerting response, as shown in Fig. 4.2 [17, 18].

It can be seen in Fig. 4.2 that the Rule-based Intrusion Detection System (RIDS) contains a pre-processing performing on raw data and transforming this data into a format that is more easily and effectively interpreted for further data processing for the purpose of intrusion detection. There are a number of different tools and methods used for pre-processing. One is feature extraction, which pulls out specified data that is significant in some particular context, to detect normal or anomalous activities.

In this regard the RIDS allows estimating continuous functions from data without mathematically specifying exactly how the output depends on the input. Thus, the respective blocks shown in Fig. 4.2 contain three main components:

- *Rule Base*: A rule can be defined as an ordered pair of strings. Set of rules that govern decisions about what is identified as normal and what is identified as known malicious activity. A rule base typically has a format of source, destination, service or action.

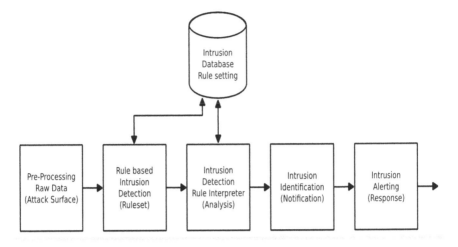

Fig. 4.2 Block diagram of a generic rule based intrusion detection system

- *Database*: Collects and organizes all activity data, and will be updated regularly. Moreover, the database contains stored known cyber threat attack intrusion event signatures and uncertain data which are compared with actual activity events for intrusion detection. In case of a security incident case an alert will be generated
- *Rule Interpreter*: Rule Interpreter: Learning kernel is based on an inference engine for decision making with regard to normal and anomalous activities in data or information that successfully match against the cyber threat attack intrusion related signatures in the database or a combination of several uncertainty sources. If anomalous activity is found, then the rule interpreter checks the rule base by comparing for instance the ordered pair of signatures of each rule until one is found which can successfully be matched against the "known" intrusion incident related signatures in the database in order to detect incident point and type of intrusion.

To achieve this goal, the block diagram model shown in Fig. 4.2 also contains an adaptive expert system (not shown in the block diagram) that solves problems by applying knowledge that has been generated based on expertise in the field of an application such as decision support to detect the source of intrusion incidents and suggest best possible prevention techniques and suitable controls for the different types of cyber threat attack intrusions. Moreover, from Fig. 4.2 it can be seen that the Intrusion Detection System (IDS) uses a security audit such as a vulnerability scan as well as alerting and reporting mechanisms. Therefore, the intrusion database has stored the "known" malicious incidents for future cyber threat attack intrusion detection. Furthermore, rules are defined and are stored in the rule set based intrusion detection engine of the system while intrusion points and types is passed to the expert system to evaluate that data with "known" malicious incidents stored in the intrusion database to detect the cyber threat attack source using a backward chaining approach. In more advanced intelligent intrusion detection system models the intelligent IDS suggests the appropriate prevention technique after detecting the cyber threat attack incident. In this regard the integrated expert system approach in the IDS permits the incorporation of human experience into the rule based intrusion detection system and then utilizes that knowledge to identify incidents that match the defined characteristics of misuse and attack. Rule based analysis as shown in the block diagram in Fig. 4.2 relies on sets of predefined rules that can be repeatedly applied to a collection of facts implemented. Facts represent conditions that describe a certain situation in the audit records or directly from system activity monitoring and rules represent heuristics that define a set of actions to be executed in a given situation and describe known intrusion scenario(s) or generic techniques. The rule that fires has identified a malicious incident and it then causes an alert that takes further action. Expert systems are able to detect intrusion incidents by encoding intrusion scenarios as a set of rules. These rules replicate the partially ordered sequence of actions that include the intrusion scenario. The identified malicious incident generates a notification feature as outcome of the intrusion detection device, shown in Fig. 4.2, which starts an alert response as an operational routine to encapsulate the identified intrusion incident. Hence, the intrusion detection architecture, shown in Fig. 4.2, is a core element in controlling the data flow between attack surfaces and mission-critical and crucial devices or components.

In addition to the described generic intrusion detection model shown in Fig. 4.2, the detection of cyber threat attacks needs an expansion of its ruleset-based approach due to the growing complexity of computer systems or networks by integrating intelligent signature detection methods. Otherwise this can result in providing invalid, unexpected, or random data which requires expanding the block "Rule based Intrusion Detection (Ruleset)". Is intrusion detection done on interfaces that cross a trust boundary, the violation refers to vulnerabilities where running software trusts data that has not been validated before crossing a boundary.

However, as reported in [19], numerous static, dynamic, and hybrid solutions are available for analyzing patterns and signatures in program codes and the behavior of program executions in order to identify the presence of malicious cyber threat attacks in the computer system or network under inspection, helping to disable them. In real time systems, which are used for mission-critical tasks, intrusion incidents can be detected through static timing analysis.

In [19], mechanisms for time-based intrusion detection are described that detect the execution of unauthorized instruction incidents in real time computer system or network environments. Such intrusion detection utilizes information obtained by static timing analysis. For real time systems, timing bounds on code sections are available as they are already determined prior to the schedule analysis. In [20], it is demonstrated how to provide micro timings for multiple granularity levels of application code. Through bound checking of these micro timings, techniques have been developed to detect intrusion incidents (1) in a self-checking manner by the application and (2) through the Operating System Scheduler (OSS), which is a novel contribution in the real time system domain.

Another important direction IDS concerns the application of Artificial Neural Network (ANN) techniques, for the anomaly and misuse detection in computer systems or networks. Artificial Neural Networks are non-linear statistical data modeling methods that try to emulate the functions of biological neural networks. ANN consists of interconnected collection of simple processing elements or artificial neurons and processes information in a connectionist approach to computation which means its treatment is to transform a set of inputs to a set of searched outputs, through a set of simple processing units, or nodes and connections between them [7]. The ability dealing with uncertainty and partially true data make ANN attractive for use in intrusion detection. In this regard some IDS have been exploited as pattern recognition technique approaches, implemented by using a feed-forward ANN that has been trained accordingly.

4.2   Intrusion Prevention

An Intrusion Prevention System (IPS) detects cyber threat attack intrusion incidents of computer systems or networks and takes immediate defensive measures. In this regard it provides additional protection over conventional firewall systems. A firewall is a security system that is capable of analyzing data or packet traffic of computer systems or networks. Therefore, it protects Information Technology (IT) systems against cyber threat attack intrusion incidents or unauthorized access. A firewall can be implemented as dedicated hardware or as a software component in public and private IT infrastructures. However, intrusion prevention differs from intrusion detection by the characteristic that intrusion prevention is some kind of a computer system or network security system with the scope detecting cyber threat attack intrusion incidents and respond to a detected cyber threat attack incident by attempting to prevent it from succeeding.

4.2.1   Intrusion Prevention System

An Intrusion Prevention System (IPS) continuously monitors computer systems or networks to detect possible malicious incidents of potentially damaging computer system or network traffic, and capturing information about them. Hence, the IPS reports these malicious incidents to intelligent algorithms to take further preventative action, such as closing access points or configuring firewalls to prevent future malicious attacks, and others. In contrast, the IDS is a passive type of security system that scans data sets or packet traffic to detect security problems and cyber threat attack incidents and reports back on detected security problems or cyber threat attack intrusion incidents, and request for further action needed. Furthermore, the IPS is also capable of detecting cyber threat attack incidents but is actively working and taking measures to protect the computer system or network. For this purpose, the IPS is installed inline directly in the legitimate data or packet traffic transmission path, and can block individual data sets or packets or interrupt and reset the connections in the incident of an intrusion alert. In today's IPS these systems work directly together with a firewall and actively influence its rules. Since the IPS works in-line, data or packet traffic analysis must be achieved in real-time. However, the IPS must not slow down the data or packet traffic stream or suspend the analysis of the data or packets due to high transmission speeds.

In order to detect anomalies or direct cyber threat attack signatures or patterns, IDS and IPS basically use the same methodological approaches. Known attack signatures or patterns can be found by comparing the analysed data or packet traffic stream by comparing them with the ones stored in a database, which refers to the "known-knowns" cybersecurity risk level of information security, as shown in Table 3.1. However, it has to be mentioned that the more extensive and up-to-date this database is, the more effective this type of detection is. In addition to this signature or pattern-based detection, additional statistical and anomaly-based methods are used. They are able to detect deviations from normal data or packet traffic streams as well as previously unknown cyber threat attack signatures or patterns, and methods, which refers to the "known-unknowns" cybersecurity risk level of cybersecurity, as shown in Table 3.1. Furthermore, the modern IPS uses more advanced methods like artificial intelligence and work partly self-learning.

4.2.2   Intrusion Prevention System Architecture

The Intrusion Prevention System Architecture (IPSA) is based on the conceptual approach that the prevention component lies in the direct communication path between the attack surface and the mission critical and crucial systems. Therefore, the IPSA actively analyzes and takes automated actions on all data or packet traffic streams that enter the computer system or network, as shown in Fig. 4.3.

The work task of the IPSA is scanning data or packet traffic streams on computer systems or networks for different cyber threat attack intrusion incidents. For this purpose IPSA performs data or packet traffic stream inspection in real time, deeply inspecting every data or packet traffic stream that travels across the computer system or network. If any malicious or suspicious data or packet traffic streams are detected, the IPSA carry out one of the following actions:

• Blocking data or traffic flows from the attack surface.
• Dropping malicious cyber threat attack protecting the critical computer systems or networks.
• Remove or replace any malicious content that remains to the computer system or network following a cyber threat attack. This is done by repackaging payloads; removing header information and removing any infected attachments from file or email servers.
• Resetting the connection.

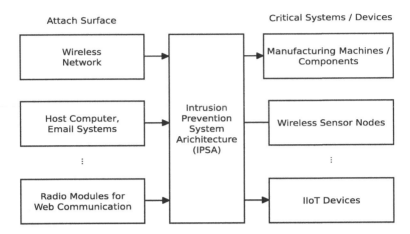

Fig. 4.3 Intrusion prevention system architecture (IPSA) lies in between the attack surface and the mission critical systems/devices

- Sending an alarm to the cybersecurity team of the public and private organization in charge.
- Terminate the Transmission Control Protocol (TCP) session that has been exploited and block offending source Internet Protocol (IP) address or user account from accessing any application, target hosts or other computer system or network resources unauthorized.
- And others.

These are essential constraints which have to be taken into account when developing an IPSA. Besides these activities, the IPSA must also detect and respond accurately, to eliminate cyber threat attack intrusion incidents and false positive detection rates, to avoid legitimate data or traffic packets misread as cyber threat attack intrusion incidents. As inline cybersecurity prevention component, the IPSA must also work efficiently to avoid degrading computer system or network performance, and must react fast because cyber threat attack intrusion incidents can happen in near real time. Moreover, IPSA solutions can be used to identify issues with corporate security policies, deterring employees and network guests from violating the rules these policies contain. In this regard the main difference between IPSA and IDSA is finally the action they take when a potential incident has been detected which result in the following characteristics:

- *Intrusion Prevention System Architecture (IPSA)*: Control access to an information and communication technology (ICT) computer system or network and protect it from abuse and cyber threat attack. These IPSs are designed to monitor intrusion data and take the necessary action to prevent a cyber threat attack from successful developing.

- *Intrusion Detection System Architecture (IDSA)*: Not designed to block cyber threat attacks. Designed to monitor the computer system or network and send alerts to computer system or network supervision intelligent algorithms if a potential cyber threat attack is detected.

Furthermore, an IPS is typically configured to use a number of different approaches to protect the computer system or network from unauthorized access. In this context, IPS solutions offer proactive prevention against some of today's most notorious computer system or network exploits. When deployed correctly, an IPS prevents severe damage from being caused by malicious or unwanted threats or brute force attacks. When deployed correctly, an IPS prevents severe damage from being caused by malicious cyber threat attacks or unwanted packets as well as brute force attacks. Several techniques are used for IPS, which can be divided into the following groups [21]:

- *IPS Stops Intrusion Attack Itself*: Examples of how this could be achieved are as follows:

  - Block access to target or possibly other likely targets from offending user account, IP address, or other intrusion attacker attributes
  - Block all access to targeted system, service, application, or other resources
  - Terminate network connection or user session that is being used for intrusion attack

- *IPS Changes Security Environment*: IPS could change configuration of other security controls to disrupt an intrusion attack. Common examples are:

  - Cause patches to be applied to a host if IPS detects that the system has vulnerabilities
  - Reconfigure a network device, e.g., firewall, router, switch, to block access by the intrusion attacker or to the target, and alter a system-based firewall to a target to block incoming attacks.

- *IPS Changes Intrusion Attack's Content*: Some IPS technologies can remove or replace malicious portions of an intrusion attack to make it benign.

  - A simple example is an IPS that removes an infected file attachment from an e-mail and then permits the cleaned email to reach its recipient
  - A more complex example is an IPS that acts as a proxy and normalizes incoming requests, which means that the proxy repackages the payloads of the requests, discarding header information. This might cause certain intrusion attacks to be discarded as part of the normalization process.

Therefore, the main task of intrusion prevention is to defend computer systems or networks by detecting a cyber threat attack and possibly repelling it. Detecting hostile cyber threat attacks depends on the number and type of appropriate actions, which can be obtained from publicly available data, found in the National Vulnerability Database (NVD), the US Government Repository of Standards Vulnerability Management Data, or the CVE database, a dictionary of publicly known information security vulnerabilities and exposures [7]. Therefore, intrusion prevention requires well-selected investigations of cyber threats because cyber threat attackers are seeking out and exploiting network, device, and application vulnerabilities to attack, causing serious problems for the computer systems or networks attacked.

Besides the foregoing mentioned it can be stated that the advancements of Next Generation Fire Walls (NGFWs) intrusion prevention and detection for computer systems or networks get closer responding to cyber threat attacks in near real time to protect the most critical and crucial data and application assets. In a more common sense NGFW is hardware and/or software based security solution used to detect and block sophisticated cyber threat attacks. They work based on security guidelines on the application layer as well as on the protocol and port checking of classic firewalls and enable data analysis at the application level. Against this background NGFW combine the functionalities of conventional firewalls which include [22]:

- *Packet-Filtering Firewalls (PFF)*: Operate at network layer (Layer 3) of the Open System Interconnection (OSI) model. Processing decisions based on network addresses, ports or protocols. This firewall is good for high performance egress filtering.
- *Quality of Service (QoS)*: Manage available network bandwidth to make sure that important network services are given priority over less important traffic, and features that are normally not available in firewalls which includes intrusion prevention options.
- *Secure Socket Layer (SSL)*: Technology responsible for data authentication and encryption for internet connections. It encrypts data being sent over the internet between two systems so that it remains private.
- *Secure Shell (SSH)*: Protocol that is frequently permitted through firewalls. Unrestricted outbound SSH is very common, especially in smaller and more technical organizations. Inbound SSH is usually restricted to one or very few servers.
- *Deep Packet Inspection (DPI)*: Information extraction (IX) or complete packet inspection (CPI) is a type of network packet filtering. It evaluates the data part and the header of a packet that is transmitted through an inspection point, weeding out any non-compliance to protocol, spam, viruses, intrusions, and any other defined criteria to block the packet from passing through the inspection point.
- *Reputation-based Malware Protection (RMP)*: Leverages the anonymous software usage patterns of users to automatically identify new cyber threats. Support vector machines can be employed on behavioural log to identify cyber threats.
- *Malware Filtering (MwF)*: Aims to stop cyber threat attack and can be based on a stochastic security game framework.
- *Application Awareness (AA)*: Represent the capacity of a computer system or network to maintain information about connected applications to optimize their operation and that of any subsystems that they run or control.

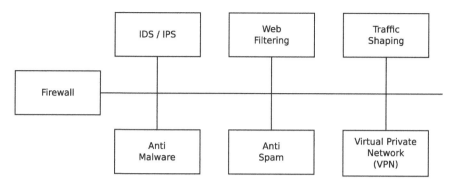

Fig. 4.4 NGFW security services platform

In this context, a NGFW represent a firewall generation that integrates intrusion detection and prevention, malware filtering, and many other security functions to allow a more advanced control of data traffic flow, as indicated in Fig. 4.4, showing the NGFW security services platform.

The blocks indicated in Fig. 4.4 have the following meaning [22]:

- *Firewall*: Providing multi-layer and protocol inspection, network segmentation, and access control.
- *Intrusion Detection and Prevention System (IDPS)*: Featuring wide range of detection techniques (ex: header-based, pattern matching, protocol-based, heuristic-based, anomaly-based), and rich customization capabilities.
- *Anti-Malware*: Providing malware protection on all webs, mail, and file transfer traffic.
- *Web Filtering*: Enforcing access to allowed web content and filtering high risk URLs such as anonymizers and known hostile addresses.
- *Anti-Spam*: Mitigating directory harvesting attacks, spam, and enforcing email policy.
- *Traffic Shaping*: Apply Quality-of-Service (QoS) to various applications' traffic such as: instant messaging (IM), web, streaming video and audio, or Peer to Peer (P2P) if allowed.
- *Virtual Private Network (VPN)*: Provide remote access and secure site-to-site interconnection over untrusted networks. Support protocols such as IPS, SSL.

Hence, if any malicious or suspicious data set or packets are detected, NGFW will carry out the following action, detection and prevention as well as reprogram or reconfigure the NGFW to prevent a similar attack occurring in the future.

4.3   Intrusion Detection and Prevention Architecture

A major challenge for organizations in today's world is on the one hand to be aware about cybersecurity needs securing mission critical and crucial data, and on the other hand to meet the resulting cybersecurity needs. Methods for logging data,

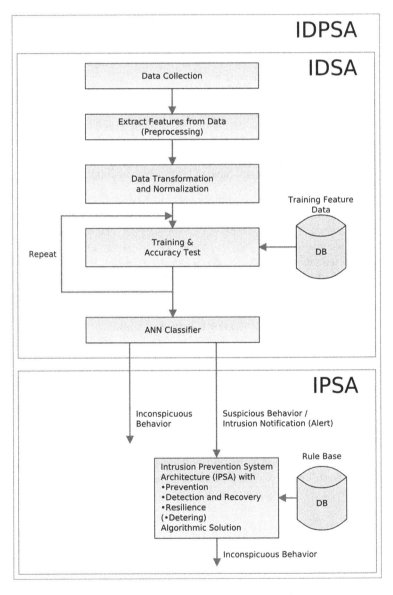

Fig. 4.5 Intrusion detection and prevention system architecture (IDPSA)

detecting intrusions, preventing intrusions have been evolving for years and are essential part of today's research [23]. Therefore, this section presents a solution to combine computer system and network based intrusion detection and prevention systems. The main activities of an Intrusion Detection and Prevention System Architecture (IDPSA) are summarized in Fig. 4.5. The Intrusion Detection System Architecture (IDSA) in Fig. 4.5 is based on the generic approach shown in Fig. 4.3, expanded by an Artificial Neural Network (ANN) classifier. The IDSA is configured in inline mode, so that data or packets are captured and in case any suspicious activity is detected by an ANN and encapsulated and alerted. ANNs are machine learning algorithms inspired by the human central nervous system. Most of them use the self-organizing map (SOM) learning algorithm, while the supervised learning algorithm is based on the perceptron approach to learn the characteristics of normal system activity and identify statistical variations from the normal trends, a research done by Fox et al. [24]. These early approaches are much more advanced ANN-based IDS today, as the architecture shown in Fig. 4.5.

Today, the backpropagation algorithm is the workhorse of training in ANNs. Executing this approach requires data gathering and pre-processing first which means that all incoming data is collected, transformed and normalized to standard entities. Thereafter, feature extraction from this data is required in which feature entities are objects of data that could be used like performance evaluation for number of data or packets transferred between computer system or network entities, delay in transfer of data or packets, number of dropped data or packets and others. Therefore, the ANN type used in IDPSA is based on the important decision matching with accuracy. In case of the IDSA in Fig. 4.5 a Feed-Forward Artificial Neural Network (FFANN) is used, which is some kind of a Multi-Layer Perceptron (MLP), consisting of an input layer with as many neurons as number of features used for classification, two hidden layers with, for example, less number of neurons and a final output layer. The goal of a FFANN is to approximate some function f^*. For example, a classifier $y = f^*(x)$ maps an input x to a category y. A FFANN defines a mapping $y = f(x;\theta)$ and learns the value of the parameters θ that result in the best function approximation.

The usage of the FFANN requires training, based on the features mentioned before. The next step after training the FFANN is to test it in place with the features assigned to normal and abnormal behavior, based on a performance metric, which describe the accuracy of the detection rate and false alarm rate of the IDSA shown in Fig. 4.5. The performance of the IDS, shown in the part IDSA in Fig. 4.5, can be calculated by the ratio of correct classification of total test data by adopting to a common model of performance measures as described in [3, 24] with the terms

- *Accuracy (A_{cc})*: Refers to the overall effectiveness of the chosen algorithm in terms introduced in Table 4.2.
- *Detection Rate (DR)*: Refers to the number of impersonation attack events detected divided by the total number of impersonation attack events in the test data set.

- *Precision (P)*: Refers to the number of impersonation attack events detected among the total number of events classified as an attack.
- *False Alarm Rate (FAR)*: Refers to the number of normal events in the test data set.
- *False Negative Rate (FNR)*: Refers to the number of attack events that are unable to be detected.
- *Score (S_C)*: Refers to the harmonic mean of *P* and *DR*.
- *Correlation Coefficient (CC)*: Represents the correlation between detected and observed data.

Accuracy is the ratio of correct detection indicated by the True Positive Rate (*TPR*) and the True Negative Rate (*TNR*), divided by the sum of True Positive Rate (*TPR*) and True Negative Rate (*TNR*) and False Positive Rate (*FPR*) and False Negative Rate (*FNR*). *TPR* represents the number of cyber threat attack intrusion events that are correctly classified as a cyber threat attack, and *TNR* is the number of cyber threat attack intrusion events, that are correctly classified as a likelihood level rare or unlikely. False Negative Rate (*FNR*) is the number of cyber threat attack intrusion events that are incorrectly classified with a likelihood of rare or unlikely, and False Positive Rate (*FPR*) is the number of cyber threat attack intrusion events that are incorrectly classified as a cyber threat attack. Finally the false alerting is the ratio of false positive divided by true negative and false positive as follows:

$$A_{CC} = \frac{TPR + TNR}{TPR + TNR + FPR + FNR}$$

The Detection Rate (*DR*) is the ratio of the number of correct detection (*TPR*), divided by the sum of true positive rate (*TPR*), and false negative rate (*FNR*) as follows:

$$DR = \frac{TPR}{TPR + FNR}$$

The Precision (*P*) is the ratio of the number of correct detection which represents the number of intrusions that are correctly classified as a cyber threat attack intrusion incident, divided by the sum of True Positive Rate (*TPR*), and False Positive Rate (*FPR*) as follows:

$$P = \frac{TPR}{TPR + FPR}$$

The false alarm rate is the ratio of the number of False Positive Rate (*FPR*), which is the number of intrusion incidents that are incorrectly classified as a cyber threat attack, divided by the sum of True Negative Rate (*TNR*), which is the number of intrusions that are correctly classified with a likelihood of level rare or unlikely, and False Positive Rate (*FPR*), which is the number of intrusions that are incorrectly classified as a cyber threat attack, as follows:

$$FAR = \frac{FPR}{TNR + FPR}$$

The false negative rate is the ratio of the number of False Negative Rate (*FNR*), which is the number of intrusions that are incorrectly classified with a likelihood of rare or unlikely, divided by the sum of False Negative Rate (*FNR*), which is the number of intrusions that are incorrectly classified with a likelihood of rare or unlikely, and True Positive Rate (*TPR*), which represents the number of intrusions that are correctly classified as a cyber threat attack, as follows:

$$FNR = \frac{FNR}{FNR + TPR}$$

The score is the ratio of the number of True Positive Rate (*TPR*), which represent the number of intrusion incidents that are correctly classified as a cyber threat attack with the weight 2, divided by the sum of True Positive Rate (*TPR*), which represents the number of intrusion incidents that are correctly classified as a cyber threat attack with the weight 2, and False Positive Rate (*FPR*), which is the number of intrusion incidents that are incorrectly classified as a cyber threat attack, and False Negative Rate (*FNR*), which is the number of intrusion incidents that are incorrectly classified with a likelihood of rare or unlikely, as follows:

$$S_C = \frac{2TPR}{2TPR + FPR + FNR}$$

The correlation coefficient, as introduced in [3], is the ratio of the number of True Positive Rate (*TPR*), which represents the number of intrusion incidents that are correctly classified as a cyber threat attack, and (*TNR*), which is the number of intrusion incidents that are correctly classified with a likelihood of level rare or unlikely, and False Positive Rate (*FPR*), which is the number of intrusion incidents that are incorrectly classified as a cyber threat attack, and False Negative Rate (*FNR*), which is the number of intrusion incidents that are incorrectly classified with a likelihood of rare or unlikely, divided by the square root of True Positive Rate (*TPR*) and False Positive Rate (*FPR*), True Positive Rate (*TPR*) and False Negative Rate, True Negative Rate (*TNR*) and False Positive Rate (*FPR*), True Negative Rate (*TNR*) and False Negative Rate (*FNR*).

$$Correlation\ Coefficient = \frac{(TPR \times TNR) - (FPR \times FNR)}{\sqrt{(TPR + FPR)(TPR + FNR)(TNR + FPR)(TNR + FNR)}}$$

In this context abnormal behavior can be received using a statistical based threshold approach. Thus, the FFANN learns through an iterative process classifying any feature into normal or abnormal classes based on the profile created by FFANN for both classes. A stable trained FFANN then recognizes and classifies the data or packets and control messages in the computer system or network under test in real time by generating a decision for normal or abnormal behavior. In case of abnormal behavior, which may result in a cyber threat attack intrusion alert of the Intrusion Detection System Architecture (IDSA), and the Intrusion Prevention System Architecture (IPSA), both parts of the overall Intrusion Detection and Prevention System Architecture (IDPSA), are immediately activated. Hence, the Intrusion Prevention System Architecture (IPSA), as indicated in Fig. 4.5, is the key element protecting data or traffic flows or packets between computer systems or networks in case of cyber threat attacks through

- *Prevention*: Technology that examines network data or traffic flow to detect and prevent vulnerability exploits
- *Detection and Recovery*: Detects response problems from vulnerability and recovers to a functional state. In this regard it is something like a timeout window in between detection and recovery
- *Resilience*: Capacity to recover quickly from a vulnerability
- *Deterring*: Some action taken that prevents people from doing something by making them afraid.

The foregoing mentioned performance metrics to measure the effectiveness of an IDSA can be divided into three classes [25, 26]:

- *Threshold Metric*: Includes features such as Classification Rate (*CR*)—ratio of correctly classified events and the total number of events—F-Measure (*FM*)—estimate of how accurate a classifier is—Cost per Example (*CPE*), and others. This metric considers whether the prediction is below a threshold, whereby the threshold lies in between 0 to 1.
- *Ranking Metric*: Include False Positive Rate (*FPR*), Detection Rate (*DR*), Precision (*P*), Area under Curve (*AuC*)—Curve used to visualize the relation between *DR* and *FPR* of a classifier, and to compare the accuracy of classifier(s). This measure is effective but has some limitations, depending on the ratio of cyber threat attacks to normal data or packet traffic events
- *Probability Metric*: Includes Root Mean Square Error (*RMSE*) and lies in the range from 0 to 1. Metric is minimized when the predicted value for each cyber threat attack class coincide with the true conditional probability of that class being normal class.

Table 4.4 Confusion Matrix

Event	Predicted cyber threat attack	Predicted normal event
Cyber threat attack	*TPR*: Intrusions successfully detected by IDS	*FNR*: Intrusion misused by IDSA classified as normal/nonintrusive
Normal event	*FPR*: Normal/nonintrusive event, wrongly classified by IDS	*TNR*: Normal/nonintrusive event, successfully classified as normal/nonintrusive by IDS

These metrics can be used to calculate the performance for the classification of results for IDS in the form of the so called confusion matrix. Since the confusion matrix refers to classification results it represents the *TPR*, *TNR*, *FPR*, and *FNR* classification results of the IDSA. In Table 4.4 the results of possibilities to classify cyber threat attack intrusion events are shown.

4.4   Intrusion Detection Capability Metric

Evaluation of cyber threat attack intrusion detection is fundamental to measure the objective effectiveness of IDSs in terms of their ability to correctly classify detected events as normal or malicious incidents. Therefore, measuring the capability of IDSs is essential in IDS research and application, because it enables the process of deciding cyber threat attack intrusion detection accuracy. In this regard measuring the cyber threat attack intrusion detection capability allows developing an IDSA that maximizes the capability metric. However, there are several metrics available, which allows measuring different aspects of an IDSA, but none of them objectively measure the cyber threat attack intrusion detection capability of an IDSA [27]. Therefore, it is difficult to determine which IDSA is better than another to objectively detect malicious intrusions in terms of only False Positive Rate (*FPR*), i.e. the probability that the IDSA outputs an alarm, when no cyber threat attack intrusion occurs, and True Positive Rate (*TPR*), i.e. the probability that the IDSA outputs an alarm when there is a cyber threat attack incident. Let's assume, IDSA1 detects 10% more cyber threat attacks than IDSA2, but IDSA2 generates 10% less false alarms. Now the question is which IDSA is better, IDSA1 or IDSA2? This raises a question: How to determine a unified objective metric that allows calculating the Intrusion Detection Capability (*CID*) of IDSAs? In [28], a metric is suggested which allow selecting the best IDSA configuration for an operational cyber threat attack intrusion detection infrastructure in public or private organization environments infrastructure. This means this metric is able to evaluate different IDSAs to choose the one which fits best by analyzing input data or packet streams as being normal or malicious and generating an alert if truly malicious. The model for cyber threat attack intrusion detection developed as shown in Fig. 4.6.

Fig. 4.6 Abstract intrusion detection model

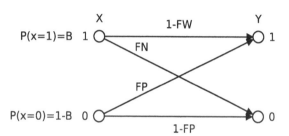

The model in Fig. 4.6 is based on the following assumptions: Let a random variable X be equal to 1 if a cyber threat attack intrusion incident happens, and X be equal to 0 in normal data or packet streams. Hence, the IDSA outputs an alert if a random variable Y equals 1 while a cyber threat attack intrusion incident happens, and if the random variable Y equals 0 it means no alert as there is no cyber threat attack intrusion incident happening. Thus, the IDSA generates an output corresponding to each input. Let's assume that data or packet streams be encoded for formal reasons as follows: $X = 1$ for malicious incidents, and $X = 0$ for normal data or packet streams, the IDS examines every packet stream to classify the output indicating a malicious incident $Y = 1$ or not $Y = 0$.

In the model shown in Fig. 4.6, $p(X = 1)$ represents the so called base rate B, which is the prior probability of a cyber threat attack intrusion to the input data examined and detected by the IDSA, the "known-knowns" (see Table 3.1).

Let's assume that a cyber threat attack intrusion incident has a probability $p(Y = 0|X = 1)$ and being considered as normal by the IDSA. This represents a False Negative Rate (*FNR*), which can be denoted by the symbol β, identifying that there is no alert A, when there is a cyber threat attack intrusion event I.

Let's assume a normal event has a probability $p(Y = 1|X = 0)$ and being misclassified as a cyber threat attack intrusion incident by the IDSA. This represents a False Positive Rate (*FPR*), which can be denoted by the symbol α, identifying that there is an alert A, when there is no cyber threat attack intrusion I.

However, evaluating an IDSA, a data set has to be given for evaluation, to run the test for being able to calculate symbols B, a, and β.

Henceforth, the model shown in Fig. 4.6, introduced in [28], is used for intrusion detection purpose from an information theoretic perspective, which finally results in an Intrusion Detection Capability (CID) metric, to select the best IDSA configuration for an operational cyber threat attack intrusion event detection infrastructure as well as to evaluate different types of IDSAs with regard to fitting the best *CID*. Therefore [28] defines the *CID* as follows: "Let X be the random variable representing the IDSA input and Y the random variable representing the IDSA output. *CID* is defined as:"

$$CID = \frac{I(X;Y)}{H(X)} = \frac{H(X) - H(X|Y)}{H(X)}$$

whereby $I(X;Y)$ represents the mutual information of X and Y, $H(X)$ is the entropy of X, and $H(X|Y)$ is the conditional entropy of X after Y is known. The mutual information measures the reduction of uncertainty of the IDSA input, knowing the IDSA output, and thus can be normalized using the entropy, i.e., the original uncertainty of the input. Hence, *CID* represents the uncertainty reduction ratio of the IDSA input, to estimate the IDSA output. Its value range is [0, 1]. Therefore, a larger *CID* value means that the respective IDSA has a better capability classifying input cyber threat attack intrusion events accurately.

4.5 Intrusion Detection and Prevention Methods

Another option is testing for stability and resiliency because the complex software systems found in today's public and private organization infrastructure environments are prone to attacks. These environments need to fully assess security to ensure a stable and resilient system. To test for stability and resiliency, several methodologies are used:

- *Functional and Performance Test*: Validates security components under valid traffic and cyber threat attack conditions
- *Impairment Test*: Validates performance when communication is impaired; typically used with delayed, dropped, or erroneous packets
- *Resiliency Test*: Validates operation under degraded or failure conditions, such as sensor failure, actuator failure etc.
- *Stress Test*: Validates system or components beyond normal operational capacity to observe how the system or components operate.

Besides these test methods other types of security testing are:

- *Access Control Test*: Ensures that the computer system or network and its application under test can only be accessed by authorized and legitimate users. The objective of this test is to assess the differentiation of the software components and ensure that the respective application implementation conforms to the security policies and protects the computer system or network from unauthorized users.
- *Ethical Hacking Test*: Person who attacks the computer system or network, mimicking the manner of cyber-hackers. The computer system or network as well as an application is attacked from within to expose security flaws and vulnerabilities, and to identify potential cyber threat attacks that malicious hackers might take advantage of.
- *Security Risk Assessment Test*: Involves the risk of the security system by reviewing and analyzing potential cyber-attack risks. This type of cyber-attack risk can be classified into high, medium and low categories based on their severity level. Thereafter, the respective mitigation strategy follows based on the security posture of the computer system and/or network application. Security audits conducted for access points, inter-network, intra-network access, and data protection are done at this level.
- *Security Scanning Test*: Enhance the scope whereby testers conduct security scans to evaluate computer system or network weaknesses. Each scan sends malicious requests to the system and testers check for behavior that could indicate security vulnerability. SQL Injection, XPath Injection, XML Bomb, Malicious Attachment, Invalid Types, Malformed XML, Cross Site Scripting etc. are some of the scans that need to run to check for vulnerabilities which are then studied at length, analyzed and then fixed.

- *Vulnerability Scanning Test*: Tests the entire computer system or network under test to detect system vulnerabilities, loopholes, and suspicious vulnerable signatures. This scan detects and classifies the system weaknesses and also predicts the effectiveness of the countermeasures that have been taken.

With regard to tests another important strategy is the Security Penetration Test (SPT) which is a simulated test that mimics a cyber threat attack intrusion incident by a hacker on the computer system or network being tested. This test aims to gather information about the computer system or network and identifying entry points into the computer system or network or application attempting a break in to determine the cyber security weakness. This type of test is like a white hat cyber-attack. When implementing the SPT, the same techniques, tools and expert knowledge are used, which are also used by cyber attackers. However, experienced penetration testers are required which use automated and manual test procedures to present realistic attack scenarios. This will become important with regard to the large number of distributed and connected computing resources in the future smart manufacturing infrastructure of Industry 4.0 where anything will be connected with everything by a network representing a cloud environment to deliver and/or exchange essential manufacturing data and/or machinery access data, which also require advanced cybersecurity methods to protect the cloud system against possible cyber threat attacks.

References

1. R. Heady, G. Luger, A.B. Maccabe, M. Servilla, The Architecture of a Network Level Intrusion Detection System, Technical Report 390-20, Department of Computer Science, University of New Mexico, 1990
2. J. Anderson, *An Introduction to Neural Networks* (MIT Press, Cambridge, MA, 1995)
3. K. Kim, M.E. Aminanto, H.C. Tanuwidjaja, *Network Intrusion Detection using Deep Learning—A Feature Learning Approach* (Springer Nature, Singapore, 2018)
4. R. Tiwari, R. Kumar, A. Bharti, J. Kishan, Intrusion detection system. Int. J. Techn. Res. Appl. **5**, 38–44 (2017)
5. B. Cappers, Interactive Visualization of Event Log for Cybersecurity, PhD Thesis, TU Eindhoven, 2018
6. S. Kumar, E.H. Stafford, A pattern matching model for misuse intrusion detection, in *Proceedings of the 17th National Computer Security Conference*, 1994, pp. 11–21
7. D.P.F. Möller, R.E. Haas, *Guide to Automotive Connectivity and Cybersecurity—Trends, Technologies, Innovations, and Applications* (Springer Publ., Cham, 2019)
8. B. Rhodes, J. Mahaffey, J. Cannady, Multiple self-organization maps for intrusion detection, in *Proceedings of the 23rd National Information Security Conference*, 2000, pp. 32–42
9. O. Depren, M. Topallar, E. Anarim, M.K. Ciliz, An intelligent intrusion detection system (IDS) for anomaly and misuse detection in computer networks. Expert Syst. Appl. **29**, 713–722 (2005)
10. V. Veeramreddy, V.V. Rama Prasad, K.M. Prasad, A review of anomaly based intrusion detection systems. Int. J. Comput. Appl. **28**(7), 26–35 (2011)
11. T. Kohonen, *Self-Organizing-Map* (Springer Publ., Berlin, 2001)

12. C. Modi, A survey of intrusion detection techniques in cloud. J. Netw. Comput. Appl. **36**(1), 42–57 (2013)
13. R. Mitchell, I.R. Chen, A survey of intrusion detection techniques for cyber-physical systems. ACM Comput. Surv. **46**(4), 1–55 (2014)
14. S. Shieh, V. Gligor, A pattern-oriented intrusion detection model and its applications, in *Proceedings of the Symposium on Security and Privacy*, 1991, pp. 327–342
15. A. Denning, An intrusion detection model. IEEE Trans. Softw. Eng. **13**, 222–232 (1967)
16. S.E. Smaha, Tools for misuse detection, in *Proceedings of the International Social Security Association*, 1993, pp. 711–716
17. T.F. Lunt, A. Tamaru, F. Gilham, R. Jagannathan, C. Jalali, P.G. Neumann, H.S. Javitz, A. Valdes, T.D. Garvey, A Real-Time Intrusion Detection Expert System (IDES), Final Technical Report SRI Project 6784, Contract No. N0003S89-C-0050, SRI Computer Science Laboratory, 1992
18. D.P.F. Möller, R.E. Haas, K.B. Akhilesh, Automotive electronics, IT, and cybersecurity, in *Proceedings of the IEEE/EIT Conference*, pp. 575–580
19. E. Karim, V.V. Proha, Cyber-physical systems security, in *Applied Cyber-Physical Systems*, ed. by S.S. Shuh, U.J. Tanik, J.N. Carbone, (Springer Publ., New York. 2014), pp. 75–84
20. C. Zimmer, B. Bhat, F. Mueller, S. Mohan, Time-based intrusion detection in applied cyber-physical systems, in *Proceedings of the 1st ACM/IEEE International Conference on Cyber-Physical Systems*, 2010, pp. 100–118
21. K. Scarfone, P. Mell, *Guide to Intrusion Detection and Prevention Systems* (National Institute of Standards and Technology (NIST), Gaithersburg, MD, 2007), pp. 800–894
22. A. Abdel-Aziz, Intrusion Detection and Response—Leveraging Next Generation FireWall Technology, SANS Institute Report, 2020
23. P. S. Krenke, A. Pal, A. Colaco (eds.), *Proceedings or the 3rd International Conference on Frontiers of Intelligent Computing: Theory and Applications* (Springer Publ., Cham, 2014)
24. K.L. Fox, R.R. Henning, J.H. Reed, R. Simonian, A neural network approach towards intrusion detection, in *13th National Computer Security Conference*, 1990, pp. 125–134
25. O.Y. Al-Jarrah, O. Alhussein, P.D. Yao, S. Muhaidat, K. Taha, K. Kim, Data randomization and cluster-based partitioning for hotnet intrusion detection. IEEE Trans. Cybernet. **46**, 1796–1806 (2015)
26. G. Kumar, Evaluation metrics for intrusion detection systems—a study. Int. J. Comput. Sci. Mobile Appl. **2**(11), 11–17 (2014)
27. R. Caruana, A. Niculescu-Mizil, Data mining in metric space: an empirical analysis of supervised learning performance criteria, in *Proceedings of the 10th ACM Symposium on Information, Computer and Communications Security*, 2004, pp. 69–78
28. G. Gu, P. Fogla, D. Dragon, W. Lee, B. Scoric, Measuring intrusion detection capability: an information-theoretic approach, in *Proceedings of the ACM Symposium on Information, Computer and Communications Security*, 2006, pp. 90–101

Chapter 5
Machine Learning and Deep Learning

5.1 Introduction to Machine Learning

Machine learning can be introduced as the ability of a machine to learn without being explicitly programmed. In this context machine learning powers many advances of public and private organizations and of modern society, from web searches to content filtering on social networks and many others. Machine learning is essentially a form of applied statistics with increased emphasis on the usage of computers to statistically estimate complicated functions and less on proving confidence intervals around these functions. Thus, machine learning can be used to identify objects in images, transcribe speech into text, match new items, select relevant results of search, and many others. Against this background, machine learning allows to tackle tasks that are too difficult to solve with traditional programming paradigms. In this context machine learning tasks are described in terms of how to process a problem to be solved. Typically a problem can represent a vector $x \in R^n$, where each entry x_i of the vector is another feature. Therefore, machine learning models can be assumed as a set of n input values $x_1, \ldots, x_n$ and associate them with an output y. These models learn a set of weights $w_1, \ldots, w_n$, and compute their output such as

$$f(x,w) = x_1 w_1 + \ldots + x_n w_n$$

Thus, many problems can be solved based on machine learning techniques. Some of the most common machine learning application domains is [1]:

- *Anomaly Detection*: Machine learning algorithm sifts through a set of events or objects, and flags some of them as being atypical or unusual (see Sects. 4.1.1–4.1.3). Machine learning algorithm is an algorithm that learns from data.

D. P. F. Möller, *Cybersecurity in Digital Transformation*, SpringerBriefs on Cyber Security Systems and Networks, https://doi.org/10.1007/978-3-030-60570-4_5

- *Classification*: Machine learning algorithm specifies which of k-categories some inputs belong to. To solve this problem learning algorithm generates a function f: $R^n \to \{1,\ldots,k\}$. For $y = f(x)$ the classification assigns an input described by vector x to a category identified by the numeric code y.
- *Denoising*: Machine learning algorithm has an input a as a corrupted problem vector $x_c \in R^n$ obtained by an unknown corruption process from an uncorrupted (regular) problem vector $x_u \in R^n$. The learning algorithm predicts the uncorrupted (regular) problem vector x_u from its corrupted problem vector x_c, or more generally predict the conditional probability distribution $p(x_u \mid x_c)$.
- *Density Estimation*: Machine learning algorithm learns a function $p(x)$: $R^n \to R$, where $p(x)$ is interpreted as a probability density function if x is continuous.
- *Imputation of Missing Values*: Machine learning algorithm solves the problem vector $x \in R^n$, but some entries x_i of x are missing. Thus the machine learning algorithm has to provide a prediction of the values of the missing entries.
- *Machine Translation*: Machine learning algorithm input consists of a sequence of symbols in some language, which has to be converted into a sequence of symbols in another language by the algorithm.
- *Regression*: Machine learning algorithm has to output a function f: $R^n \to R$.
- *Synthesis and Sampling*: Machine learning algorithm generates new examples that are similar to those in training data.
- *Transcription*: Machine learning algorithm observes an unstructured representation of some kind of data and transcribes it into discrete, textual form.
- And many others.

One of the earliest cyber threat attack problems solved by machine learning was spam detection making use of spam filters which create rules based on machine learning algorithms. The approach solving the spam problem is learning spam filters to recognize junk mails and phishing messages (see Chap. 2) by analyzing rules across a huge number of computer systems. In addition to spam detection, social media websites are using machine learning as a way to identify and filter abuse.

5.2   Types of Machine Learning

Machine learning techniques can be divided into several types, called predictive or supervised learning, descriptive or unsupervised learning, as well as reinforcement learning, as illustrated in Fig. 5.1.

The goal of first type of machine learning, predictive of supervised learning, is to learn a mapping from inputs x to outputs y, given a labeled set of input-output

$$D = \left\{x_i . y_i\right\}_{i=1}^{N}$$

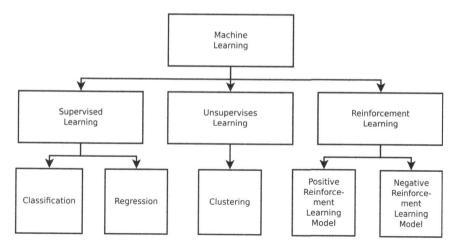

Fig. 5.1 Machine learning techniques

D is called a training set, and N is the number of training samples. In the simplest setting, each training input x is a D-dimensional vector of numbers, called features, attributes or covariates, which often is stored in a $N = D$ design matrix. However, x_i could also be a complex structured object, such as an image, a sentence, a time series, a graph, and many others. Output y can in principle be anything, but most machine learning methods assume that y_i is a categorical or nominal variable from some finite set

$$y_i \in \{I,...,Z\}$$

or a real value scalar. In case y_i is categorical, the machine learning method is called classification or pattern recognition, and if y_i is real-valued, the machine learning method is called regression [2].

A second type of machine learning is called descriptive or unsupervised learning. The machine learning goal, is finding interesting patterns in the data, which is also called knowledge discovery. Unsupervised machine learning algorithms using datasets containing features to learn useful properties of the structure of dataset.

A third type of machine learning is reinforcement learning. Reinforcement machine learning algorithms interact with an environment, so there is a feedback loop between machine learning systems and human experiences. Method is useful for learning how to act or behave for given occasional reward or punishment signals.

5.2.1   Comparison of Machine Learning Methods

The different types of machine learning partially perform complex processing tasks, based on different kinds of algorithms. Hence, each of them have advantages and disadvantages in regard to the application domain they are used for. In Table 5.1 a comparison is given for supervised versus unsupervised and reinforcement machine learning, based on same evaluation criteria to identify respective pros and cons of machine learning models.

From Table 5.1 it can be concluded in:

- *Supervised Machine Learning*: Learning model learns from a labeled data set with guidance.
- *Unsupervised Machine Learning*: Learning model is based on training based on unlabeled data without any guidance.
- *Reinforcement Machine Learning*: Agent interacts with its environment performing actions and learns from errors or rewards.

Table 5.1   Supervised learning versus unsupervised and reinforcement learning

Evaluation criteria	Supervised machine learning	Unsupervised machine learning	Reinforcement machine learning
Machine learning model	Learn on labeled data set with guidance	Training on unlabeled data without any guidance	Agent interacts with environment performing actions and learn from errors or rewards
Solved problem type	Regression and classification	Association and clustering	Reward-based
Required data type	Labeled data	Unlabeled data	No predefined data
Method of training	External supervision	No supervision	No supervision
Method of machine learning	Maps labeled inputs to known outputs	Understands pattern and discovers output	Follow trial and error method
Computational complexity	Computationally simple	Computationally complex	Depends on usefulness sample [17]
Accuracy	Highly accurate and trustworthy method	Less accurate and trustworthy method	Uncertainty reduces predictability
Best suited for	Interactive software system or applications	Automation and classification	Support and work in AI, human interaction
Cybersecurity applications	Finds functions or models that explain completely labeled data sets to detect intrusion, anomalies, and malware	Finds patterns, structures, knowledge in unlabeled data sets to detect intrusion, anomalies, DDoS attacks, and unauthorized access	Labeling data sets during acquisition to detect intrusions of cyber threat attacks

In machine learning methods applied to defend cyber threat attacks, machine learning can quickly scanning large amounts of data and analyzing it using statistics. Hence, machine learning is a powerful method used to cyber secures computer systems or network, as illustrated in Table 5.1.

5.3   Machine Learning and Intrusion Detection

Machine learning models can be used to automatically detect patterns in data which can be compared with measured data to predict whether these data are usual as expected or unusual data, indicating that an incident may happen. This incident can result in an impact on the usual operational state of the computer system or network observed which depicts a certain degree of uncertainty. Hence, machine learning models can analyze cyber threats and respond to cyber threat attacks and security incidents quickly in an automated way. In this context machine learning is a method to deal with uncertainty to gain clear decision making. Hence machine learning is an important method to perform decision making under uncertainty as it is the case in cyber threat attacks detection [3]. The detection of cyber threats based on machine learning lies in the intrinsic methodology of machine learning with its two phases of training and testing, with the following steps [4]:

- Identify class attributes and classes from training data or patterns
- Identify a subset of the attributes necessary for classification
- Learn model using training data or patterns
- Use trained model to classify the unknown data or patterns

Against this background, the public and private organizations' cybersecurity measures are used to detect and prevent cyber threat attacks in data or patterns. The concepts used can be broadly categorized into misuse detection and anomaly detection. While misuse detection methods are intended to recognize known patterns described by rules, anomaly detection focusses on detecting unusual activity patterns in observed data [5–7]. In case of misuse detection, in the training phase each misuse class is learned by using appropriate misuse scenarios from the training set (see Sect. 4.1.2). In the test phase, new data or patterns received from the system under test run through the model and the actual scenario is classified whether it belongs to one of the misuse classes or not. In case the scenario does not belong to any of the misuse classes from the training set, it is classified as normal. In the case of anomaly detection scenarios (see Sect. 4.1.1), the normal traffic pattern is defined in the training phase. In the test phase, the learned model is applied to new data or patterns received from the system under test, and building on that every scenario in the test set is classified as either normal or anomalous. The majority of the state-of-the-art cyber threat methods to public and private organizations are misuse detection due to their reliance on rule sets. Rule-based solutions can be divided into blacklist and whitelist-based approaches. A blacklist is a list of discrete entities that have been previously determined to be associated with malicious activity. A whitelist is a list of discrete entities, such as hosts, email addresses, network port numbers, run-time processes, or applications that are authorized to be present or active on a computer system or network according to a well-defined baseline. Besides blacklist and whitelist methods, graylist methods exist. A graylist is a list of discrete entities that have not yet been established as benign or malicious. Blacklist-based methods can be refined into signature-based and heuristic-based approaches to compare if a match is found which indicates a cyber threat attack incident. As described in [8], signature-based approaches allow detecting cyber threats based on specific threat patterns, for instance malicious byte sequences, while heuristic methods allow for the detection of unknown cyber threats based on an expert-based probabilistic rule set that describes malicious indicators. Although heuristic approaches often complement signature-based solutions, a major drawback is their susceptibility to high false positive rates. However, blacklist-based detection methods should be automatically updated each day and the detection should be in real time. White-list based approaches usually include policies which allow for the detection of cyber threat attacks based on the deviation from a pre-defined negative baseline configuration, for instance IP whitelists. Thus, the application of whitelist software prevents installation or execution of any application that is not specifically authorized for use on a particular host. This mitigates multiple categories of threats, including malware and other unauthorized software. An overview of state-of-the art misuse and anomaly-based detection methods is provided by the work of Modi [9] and Mitchell [10]. In this context machine learning provides methods to automatically infer generalized data models based on patterns identified in data. In supervised learning, the categorized and labeled training data feed into classification or regression models during the training phase. Thus, machine learning methods are becoming an important issue in intrusion detection. A scenario for machine learning in intrusion detection, illustrating the capability to monitor and protect an Industry 4.0 (I4.0) intelligent manufacturing environment against cyber threat attacks, is shown in Fig. 5.2. Machine learning in the scenario in Fig. 5.2 is based on data exploration to learn about regular and irregular behavior according to how I4.0 intelligent manufacturing systems and networks interact with one another within the I4.0 intelligent environment to predict malicious cyber threat attacks at early stages and output alerting, or to detect malicious cyber threat attacks and decide how to prevent them. This learning model can be introduced as a learning approach based on experience.

Moreover, machine learning is an additional value capable to detect known and new cyber threat attack types. The latter was introduced as unknowns in Sect. 3.3. In this context, machine learning algorithms compare whether identified cyber threat attacks belong to the attack category known-knowns or not. In case of not known cyber threat attack types, based on the attack category known unknowns as illustrated in Table 3.1, the detected cyber threat attack is automatically output as a new cyber threat attack of the attack category known-unknowns which refers to the cybersecurity protection issue. This learning model can be introduced as machine learning based on experience. However, this requires that the digital transformation based intelligent environment must have on the one hand a secure communication between its system devices and networks, and on the other hand a cybersecurity-based

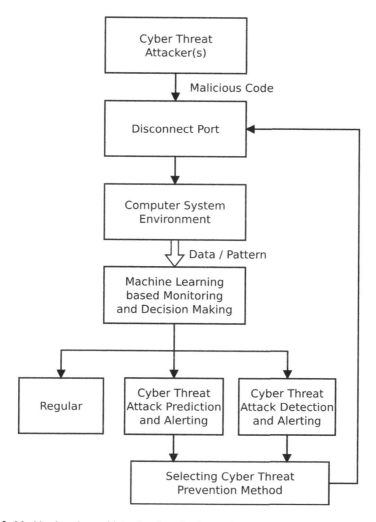

Fig. 5.2 Machine learning and intrusion detection interaction

intrinsic intelligence with focus on the capabilities of machine learning methods, especially in the attack category of known-unknown cyber threat attacks.

With regard to practicability, for instance in misuse intrusion detection, a rule based approach can be used based on association rules in the simple form

$$IF\ A\ AND\ B\ THEN\ C$$

which describes the relationship that when *A* and B are present then *C* is present too. For a real world problem this association rule can be

$$IF\ service\ request\ AND\ data\ set\ AND\ port\ address\ THEN\ attack\ type\ no.XY$$

Thus an essential initial step is finding associative rules with high confidence. Against this background, association rule mining helps to discover that the rules explain the relationships clearly. This approach is helpful to develop attack signatures for machine learning. The intrusion detection approach is supported by a signature-based machine learning method to compare each data set to all rules which describe the respective signatures, for detecting a cyber threat attack. Besides signature-based machine learning, clustering of rules is another option. In this regard association rule clustering minimizes the number of comparisons necessary to determine which rules are triggered by given input data sets [4]. In case of association rule algorithm for machine learning based intrusion detection system, attack category known-unknowns is the underlying cybersecurity approach to investigate the relationship among the various variables in the training data sets.

Let U, V, W be variables in a data set. The association rule algorithms investigate the relationship between the variables to identify their correlations and hence build a model which is used to predict the class of new samples that co-exist with cyber-attacks [11]. In [12], an association rule algorithm is discussed which shows a good performance in intrusion detection. Another promising approach in machine learning and intrusion detection combines the outputs of classification methods to generate a collective output and enhance classification performance. In this context, ensemble learning is a learning method that tries to combine heterogeneous and homogeneous multi-classifiers to obtain good classification results [13, 14]. The ensemble learning method uses several machine learning methods to reduce variance and is robust to over-fitting [15]. In Table 5.2 the potential for two machine learning methods for securing intelligent environments is summarized.

5.4   Introduction to Deep Learning

Deep Learning (DL) is a subset of machine learning that uses algorithms to imitate the human brain as faithfully as possible. Hence, deep learning is a method composed of multiple processing layers to discover representations of data with multiple levels of features. These features are automatically discovered and they are composed together in various levels to produce outputs. Each level represents abstract

Table 5.2 Machine learning models for securing intelligent environment

Machine learning	Working principle	Advantages	Disadvantages	Applications
Associated rule-based learning	Studies relationship between variables in training data sets	Simple algorithms, easy to use	Time complexity of algorithm is high	Intrusion detection
Ensemble learning	Combines concepts of different classification methods	Robust algorithm, adapt better as single classifiers	Time complexity of algorithm is high	Intrusion detection, anomaly detection, mal-ware detection

features that are discovered from the features in the previous level. This allows deep learning to be obtained by computing simpler, but non-linear models that each transforms the representation at one level, for instance raw input data, into a representation at a higher, slightly more abstract level to help solving real world problems. Deep learning models have a huge number of parameters but acquiring enough labeled data to train deep learning models is difficult. To overcome the problem of labeled training data, one can focus on unsupervised learning. The most natural way to perform this is to use generative models such as directed, undirected and mixed [2].

5.4.1   Deep Learning Methods Used in Cybersecurity

Deep learning is applicable for cybersecurity. The Deep Belief Network (DBN) is a probabilistic generative model consisting of multiple layers of stochastic and hidden variables. In their study Ding et al. [16] applies Deep Belief Nets to detect malware. Restricted Boltzmann Machine (RBM) and Deep Belief Network (DBN) are interrelated because composing and stacking a number of RBMs enables hidden layers to train data effectively through activations of one RBM for further training stages relevant to network anomaly detection based on experiments showing the feasibility of the deep learning approach to network traffic analysis. Another type of Deep Learning Networks is the Generative Adversarial Network (GAN), shown in Fig. 5.3, representing a framework for estimating generative models through an adversarial process in which simultaneously two models are trained, a generative model that captures the data set distribution, and a discriminative model that estimates the probability that a sample came from the training data sets rather than the generative model. This model was developed by Goodfellow et al. [17], in which the training procedure for the generative model is to maximize the probability of discriminative model making a mistake. The model framework corresponds to a minimax two player-game. In the space of arbitrary functions of the generative model and the discriminative model, a unique solution exists, with the generative model recovering the training data set distribution and the discriminative model to 0.5 everywhere. In case the generative model and the discriminative model are defined

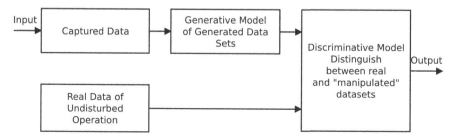

Fig. 5.3 Generative adversarial network

by multilayer perceptron's, the entire system can be trained by Backpropagation (BP). As described in [17]. The algorithm back propagates derivatives through generative processes using the observation that

$$\min_{\sigma \to 0} \nabla_x E_{0 \sim N(0, \sigma_2 I)} f\left(x + \right) = \nabla_x f\left(x\right).$$

When training has finished, the generator is capable of generating new data that is not distinguishable.

A recently published book that bridges the areas of Deep Learning and Cybersecurity, providing Deep Learning tools and frameworks to allow users to quickly develop workable and advanced prototypes, is referenced in [18]. It also introduces recent advances in the fields of intrusion detection, malicious code analysis, and forensic identification and shows how deep learning methods can be used to advance cybersecurity objectives, including detection, modeling, monitoring and analysis of as well as defense against various threats. In [19], the performance of Deep Neural Network (DNN) for cybersecurity use cases is evaluated including Android malware classification, incident detection, and fraud detection.

References

1. I. Goodfellow, Y. Bengio, A. Courvill, *Deep Learning* (MIT Press, Cambridge, MA, 2016)
2. K.P. Murphy, *Machine Learning—A Probabilistic Perspective* (MIT Press, Cambridge, MA, 2012)
3. T. Thomas, P. Vjayaraghavan, A. Emmanuel, *Machine Learning Approaches in Cyber Security Analytics* (Springer Publ., Singapore, 2020)
4. A.L. Buczak, E. Guven, A survey of data mining and machine learning methods for cybersecurity intrusion detection. IEEE Commun. Surv. Tutorial **18**(2), 1153–1176 (2016)
5. L. Potnoy, E. Eskin, S. Stolfo, Intrusion detection with unlabeled data using clustering, in *Proceedings of the ACM Workshop on Data Mining Applied to Security*, 2001, pp. 5–8
6. E. Eskin, A. Arnold, M. Prerau, L. Portnoy, S. Stolfo, A geometric framework for unsupervised anomaly detection, in *Applications of Data Mining in Computer Security*, ed. by D. Barbara, S. Jajodia, (Springer Publ., Boston, MA, 2002), pp. 77–101
7. K. Leung, C. Leckie, Unsupervised anomaly detection in network intrusion detection using clusters, in *Proceedings of the 28th Australasian Conference on Computer Science*, vol. 38, 2005, pp. 333–342
8. P. Duessel, Detection of Unknown Cyber Attacks Using Convolution Kernels over Attributed Language Models, PhD Thesis, University of Bonn, 2018
9. C. Modi, D.J. Patel, B. Borisaniya, H. Patel, A. Patel, M. Rajarajan, A survey of intrusion detection techniques in cloud. J. Network Comput. Appl. **36**(1), 42–57 (2013)
10. R. Mitchell, I.R. Chen, A survey of intrusion detection techniques for cyber-physical systems. ACM Comput. Surv. **46**(4), 55.1–55.29 (2014)
11. M.A. Al-Garadi, A. Mohamed, A. Al-Ali, X. Du, M. Guizari, *A Survey of Machine and Deep Learning Methods for Internet of Things Security* (Cornell University, Ithaca, NY, 2018) arXiv:1807.11023
12. A. Tajbakhsh, M. Rahmati, A. Mirzaei, Intrusion detection using fuzzy association rules. Appl. Soft Comput. **9**(2), 462–469 (2009)

13. M. Wozniak, M. Grana, E. Corchado, A survey of multiple classifier systems as hybrid systems. Inf. Fus. **16**, 3–17 (2014)
14. P. Domingos, A few useful things to know about machine learning. Commun. ACM **55**(10), 78–87 (2012)
15. C. Zhang, Y. Ma, *Ensemble Machine Learning Methods and Applications* (Springer Publ., Boston, MA, 2012)
16. Y. Ding, S. Chen, J. Xu, Application of Deep Belief Networks, for Opcode based Malware Detection, In: International Joint Conference of Neural Networks, pp. 3901–3908 (2016)
17. I. Goodfellow, J. Pouget-Abadie, M. Mirza, B. Xu, D.W. Farley, S. Ozair, A. Couville, Y. Bengio, Generative adversarial nets, in *Advances in Neural Information Processing Systems*, (MIT Press, Cambridge, MA, 2014), pp. 2672–2680
18. M. Alazab, M. J. Tang (eds.), *Deep Learning Applications for Cybersecurity* (Springer Publ., Cham, 2019)
19. R. Vinayakumar, H.B. Barathi Ganesh, P. Prabaharan, M. Anand Kumar, K.P. Soman, Deep-Net: Deep Neural Network for Cybersecurity Use Cases. https://arxiv.org/ftp/arxiv/papers/1812/1812.03519.pdf

Chapter 6
Attack Models and Scenarios

6.1 Introduction

Cyber threat attack is a critical and sensitive issue in the era of digital transformation. The impact and prevalence of cyber threat attacks has grown, and need action toward computer system and network security to better protect them. Cyber threat defense strategies, such as intrusion detection and prevention (see Chap. 4), strict firewall policies, penetration tests, and access controls, are some common defense approaches. Hence, cybersecurity dependence of computer systems and networks has driven the demand for pre-emptive cyber threat analysis to help in early discovery of potential cyber threat attacks or vulnerabilities. One reason for this is that white-hat cyber hackers actually attack computer systems or networks to discover vulnerabilities. Unfortunately, currently there are no standard methods for measuring the effectiveness of cyber threat risks and no standardized effective defense strategies. However, a number of options is available based on which a strategy can be accumulated to defend against cyber threat attacks (see Chap. 3). Thus, public and private organizations put enormous efforts to secure their data against cyber threat attacks. They use various types of Threat Intelligence Management Platform (TIMP) framework tools and techniques to keep their organization's daily work possible and secure it, while cyber threat attackers are trying to breach security and infiltrate malicious software to access valuable data. However, the cyber threat attack situation is getting more and more worse, because of new types of malware emerging to attack computer systems and networks. Thus it is important to gain knowledge to understand cyber threat attacks both, before and after they happen, in order to provide better insight into possible cyber threat attack scenarios. In this context cyber threat attacks refer to cyber adversaries who attempt unauthorized access to computer systems or networks using a data communications pathway by making use of various Techniques, Tactics, or Exploits (TTE), to adversely affect computer systems and networks in their direction, or steal or manipulate valuable as

© The Author(s), under exclusive license to Springer Nature Switzerland AG 2020
D. P. F. Möller, *Cybersecurity in Digital Transformation*, SpringerBriefs on
Cyber Security Systems and Networks, https://doi.org/10.1007/978-3-030-60570-4_6

well as sensitive data. These exploits can be directed through various conducts, for instance from remote locations by unknown persons using the Internet. Hence, cybercriminal attacks are facilitated by or committed by using computer systems, networks, smart hardware devices, and many others, where they are agents, facilitators, or targets of the cyber-crime [1]. Thus, to protect computer systems or networks against cybercriminal attacks, it is necessary to create a secure cyber-barrier around computer systems or networks, which require a systemic approach of knowledge in cybersecurity, This includes on the one hand the description of types of possible cyber threat attacks, which may happen, and on the second hand the development of adversary models for better understanding the scope of possible cyber threat attack problems and its intrinsic risk.

The U.S. Government Accountability Office (GOA) has published the "Guide to Supervisory Control and Data Acquisition (SCADA) and Industrial Control System Security (ICSS) draft" [2, 3], which provides a description of various cyber threats. From this draft it can be seen, that there is no universal cybersecurity solution available, with regard to the manifold of cyber threat adversary possibilities and their impacts. Henceforth, profiling cybercriminal attackers is required, because profiling allows getting more knowledge about their

- Different skills
- Personality traits
- Methods of computer criminals operation
- And many others

which help developing defense methods to protect computer systems or networks against cyber-criminal attitudes. Like in traditional crimes, it is important understanding the unknown attitude by answering questions such as

- What motivates a cyber-criminal attacker to get involved in cyber-crime?
- What keeps a cyber-criminal attacker in his cyber-criminal behavior?
- How to choose a cyber-criminal attacker and his targets?
- And many others.

However, there is no simple answer possible as it is also the case in traditional crimes. Thus, profiling cybercriminal cyber threat attackers can help to draft a picture from the puzzles gathered due to the lack of reliable data, which hinders efforts to create substantive profiles of people behind cybercriminal attacks. Therefore, superior national and international data exchange, with regard to known attacks while identified, categorized, and documented cyber threat attacks, stored in secure databases, accessible through secure access keys, would be an essential and important initiative, based on which profiling can be executed. This can lead to the integration of cyber-threat attack models with the respective scenarios used by cyber-attackers, which may lead to a more effective approach supporting intrusion detection and prevention system design. Furthermore, cyber threat attack models provide more insight into computer system or networks vulnerability, which in turn can be used to protect computer systems or networks against future cyber threat attacks. Thus, the utilization of cyber-threat attack modeling techniques provide an advanced method, which can also help during an ongoing cyber threat attack incident to identify the cyber-attacker(s) and the target of his/her cyber threat attack.

6.2 Attack Models and Scenarios

Accessibility and connectivity are two key issues in the digital transformation era, but this comes with a number of unprecedented potential risks like valuable data being stolen, losing privacy or identity, getting infected by malware, and many others. Therefore, computer systems or networks can get infected in the cyber space of digital transformation. Hence, research on cyber threat attack analysis is required to gain knowledge about the nature of cyber threat attackers' profiles, their motivation, security weaknesses of the computer systems or networks targeted to mitigating future attacks. Developing attack models of potential cyber threat attacks are an essential knowledge base for designing effective intrusion detection and prevention system models, by which adversary profiling can be executed and security enhanced.

An adversary is a Cyber Threat Attacker (CTA) or a set of cyber threat attackers, who attack a target after analyzing their capability against the target. After analyzing the capability of the target, the adversary may find that he has more capability than the targeted object of attack to attack or not. This approach is essential when dealing with more advanced cyber-attackers who have already gained some control of targeted computer systems or networks. Let's assume the adversary has also analyzed the infrastructure of the targeted resources to take over command and control (C2) of any targeted object. Against this background, the profiling adversary's technical and behavior patterns are gained to make assumptions about a cyber space acting adversary. In this context attacker models and scenarios help profiling cyber-criminal adversaries. Profiling needs collecting knowledge about potential adversaries which includes asking questions to be answered such as:

• What is the adversary's objective?
• What is the adversary's goal?
• What may be the adversaries preferred attack method to achieve his goal?
• And many more.

In this context adversary profiling depicts the attack potential or the attack risk as a measure of a minimum effort to be expended in an attack to become successful with the cyber-attack space defined by attacker's disclosure, knowledge, resources, and others. Hence, cyber threat attack modeling techniques are important to understand, explore, and validate security threats in the cyber world of the digital transformation [4]. Today, a number of cyber threat attack modeling techniques exist and are used to analyze cyber threat attacks such as attack graph or attack tree [5, 6] cyber threat attack vector [7], cyber threat attack surface [8], diamond model [9], Open Web Application Security Project (OWASP) cyber threat model [4], kill chain [10, 11], and many others. The cyber threat attack modeling techniques diamond model [9], cyber-attack kill chain [10, 11], and attack graph [5, 6] for cyber threat attack modeling are discussed in detail in [12]. The OWASP Automated Threat Handbook [13] currently describes 20 cyber threat incidents. OWASP currently work on a Top 10 publication list, describing the ten most significant classes of application vulnerabilities [14], whereby each vulnerability includes two threat modeling constructs: threat agents—types of threat actors which could exploit the vulnerability, and attack vectors descriptions of how the vulnerability could be exploited, in effect, descriptions of either threat events or fragments of threat scenarios.

Cyber threat attack modeling is the process of developing a representation of adversarial cyber threats with regard to possible symptoms of used cyber threats, scenarios, and specific incidents in the cyber space of digital transformation, sources, targeted sectors, and many others. Hence, the development of adequate adversaries' attack models is to create substantive profiles of adversaries behind cyber-criminal attacks and their intended attack scenarios with specific values such as adversary expertise, adversary resources, adversary motivation, cyber threat attack patterns, cyber threat attack incidents, and others. This requires specifying an Adversary Attack Behavior Model (AABM) which can be interpreted as a hint that cyber threat attackers may use distinct paths or alternative approaches to reach their cyber threat attack targets.

As mentioned in [15] the adversary profile depicts the attack potential as a measure of the minimum effort to be expended in an attack to be successful. In ISO/IEC 15408:2009 the attack potential is defined as "measure of the effort to be expected in attacking a Target of Evaluation (TOE), expressed in terms of an adversary's expertise, resources and motivation". Besides this, ISO/IEC 15408-1:2009 gives guidelines for specification of Security Targets (ST) and provides a description of the organization of components throughout the model. The standard is being replaced by ISO/IEC AWI 15408-1. In this context adversaries attack corresponds to the effort required creating and carrying out the targeted attack objectives. In this context adversaries' attacks correspond to the effort required creating and carrying out the targeted attack objectives. Thus knowledge about the cyber-attack space based on knowledge of objectives of the cyber threat attacker is important in attack scenario analysis for adversary attack modeling. Scenario analysis requires on the one hand the identification of the attackers' intentions with regard to the goals and tasks of the possible attacks to the computer system or network and the identification of possible countermeasures by an intrusion detection and prevention system to the identified attacks. In this context the attack space is based on a-priori knowledge of the target available to the cyber threat attacker, disclosure resources enabling the attacker to obtain target information during the cyber threat attack, and disruption resources to affect the target operation used by the cyber threat attacker, as reported in [16]. However, this requires that a security scenario attack analysis has enough information about the computer system or network and its environment to allow validation of the computer systems' or networks' security requirements with respect to particular attacks. In [17] a security scenario attack is defined as an attack situation describing the targeted computer system or network and the secure capabilities as well as possible attackers and their goals, to identify the security capabilities of the targeted computer system preventing the attackers' goals. This requires identifying the intentions of possible attackers of goals and tasks, and identifying possible countermeasures such as identifying capabilities of attackers. Also high-level cyber-treat attack scenario descriptions can be used to describe cyber attacker's behavior or specific behavior can be modeled [18, 19].

6.3 Adversary Behavior Modeling

Modeling adversary behavior is an integral concept in cyber-crime defense, because an adversary behavior model represents a formalization of cyber-crime attackers' behavior to computer systems or networks, allowing to create adversary behavior simulation to measure the effects on computer system or network configurations security. However, such a model must contain an analysis of adversary behavior and incorporate knowledge to defend the targeted computer system or network. Therefore, depending on how complete this formalization is, the adversary behavior model is based on algorithms or may simply be a series of statements with regards to capabilities and goals. Thus, cybersecurity methods currently utilize adversary behavior models to verify data flows, data packets, protocols, and others with regard to malicious incidents. Another important scope is digital forensics (see Sect. 3.4), an actual field that benefits from the use of adversary models to prove that a forensic process is forensically sound [20]. McKemmish [21] defined forensic soundness in the context of digital evidence as the combination of four criteria: meaning, errors, transparency and experience. Hence, forensic soundness is integral to the admissibility of evidence, and is analogous to the aim of maintaining data security.

Adversary behavior models are an approach modeling possible adversaries' attacks to a computer system or network based on the perspective of either a cyber threat attack defender using an asset-centric threat model or an cyber threat attacker, using the intend of an attacker-centric threat model [22, 23]. Logically, attacker-centric threat models are the most closely related to adversary models, but the primary difference is that attacker-centric threat models are intended to model distinct attacks in detail for instance, steps required to perform a spoofing attack on a computer system. These attacks should be modeled with great detail to allow developers of a defense system to reinforce the defense system from such threats. Thus, adversary behavior models represent a complete attack scenario with regard to assumptions, capabilities, and goals. Hence, adversary behavior models represent a more general approach to model attacks on a computer system and, as an outcome, consider the application of cyber threat models and adversary behavior models to be distinct [18, 19]. Networks can also be modeled in various levels of detail, for instance from complex packet-level descriptions to less detailed network terrain descriptions [24, 25].

6.3.1 Adversary Attack Behavior Modeling

Adversary's cyber threat types and their impacts, and adversary model components are general elements to describe an adversary behavior. Let's assume the adversary model is based on a-priori knowledge of the targeted computer system available to the adversary, representing the core knowledge of the Adversary Attack Behavior Model (AABM) which has the general goal of supporting and analyzing how to make an attack scenario not successful. Moreover, the adversary has gained the most important a-priori knowledge of the targeted computer system which he tries to attack ($KTCS_A$) consisting overall knowledge of targeted computer system network ($KTCSN$), and DAT_i as data available at attack time i to the adversary. Thus, adversary total a-priory target computer system knowledge yield

$$KTCS_A = (KTCSN, DAT_i)$$

Let's assume that the resulting adversaries attack policy can be described by

$$AAP = (KTCS_A, PDCTCS_A)$$

with AAP as adversaries attack policy, $KTCS_A$ as advisories total a-priori targeted computer system knowledge, and probability of data corruption in the targeted computer system ($PDCTCS_A$). In this context, profiling an adversary is essential to depict the attack potential and the attack risk as a measure of a minimum effort to be expended in an adversary attack to be successful. Based on these assumptions an AABM can be developed making use of essential characteristics in a respective adversary attack model creation:

- Cyber threat attack defenders knowledge of adversaries' capabilities ($ADKAC$) which has to be taken into account for classification of adversaries' capabilities such as skills (CACS), adversaries' goals such as intentions (AGI), and assumptions with regard to possible adversaries' profiles for attacking ($PAPA$) a computer system, which ultimately gives insight into adversaries' attack policies (AAP).
- Cyber threat attack defender probability of countermeasures ($ADPIC$) of computer system infrastructure to identified attacks.

which result in a scenario-based adversary attack defending model ($AADM$)

$$AADM = (ADKAC, ADPIC)$$

with

$$ADKAC = (CACS, ASK, PAPA)$$

and

$$AADM = CACS, ASK, PAPA, ADPIC$$

This finally results in the balance of power of adversaries and defenders as final outcome

$$AAP = AADM$$

Building on these generic assumptions, the methodological approach of an AAM with regard to the balance of power of adversaries and defenders has been derived based on adversary's attack policy, and the respective adversary attack defending model. The information and data for modeling one can make use of known security standards, for instance ISO/IEC 15408:2009, ISO/IEC 18045, ISO/IEC 27000:2012, ISO/IEC 17799:2005, NIST SP-800:30, and others, and security dictionaries, for instance Common Vulnerabilities and Exposures (CVE), CAPEC™, OWASP, Comprehensive Lightweight Application Security Process (CLASP), and others. For instance, ISO/IEC 15408:2009 defines the attack potential as "measure of the effort to be expected in attacking a Target of Evaluation (*TOE*), expressed in terms of an adversary's expertise, resources and motivation". Besides this, as mentioned in Sect. 6.2, ISO/IEC 15408-1:2009 give guidelines for specification of Security Targets (*ST*) and provides a description of the organization of components throughout the model. The standard is being replaced by ISO/IEC AWI 15408-1. Furthermore, the U.S. governmental report [26] mentions that a more fundamentally secure cyber ecosystem can help tip the balance toward those protecting networks and away from malicious cyber actors. The PhD in [27] show the obligation for cyber health and the capability approach to wellbeing is investigated which is a good approach developing adversary models.

6.3.2   Adversary Cyber Threat Simulation Modeling

Let's assume adversary attack is applied with regard to knowledge gained of the targeted system by the adversary at a given time which corresponds to the adversary's assumptions, capabilities, and goals (ACG), resulting in adversary action A. A can be defined as tuple $a = (s, t, e)$, where $s \in S$ is a source node, $t \in T$ is a target node, and $e \in E$ is an exposure such that S takes action on t by exploiting exposure e. Adversary's preferences are known which determines the probabilities for selection $a \in A$ which result in attack paths in the output of the simulation model. Furthermore, adversary's current intent due to his gained knowledge K at a given point in time results in the set of exposures E. However, the adversary's knowledge is a complex data structure which represents the target object attributes uncovered by the adversary throughout the attack. A discussion of attacker knowledge constructs can be found in [28]. The adversary's preference can be determined by $a \in A$ which results in the sequence of attack actions

$$\{a_1, a_2, \ldots, a_n\}$$

taken by the adversary's attack during a cyber threat attack scenario representing an attack path. The opportunities an adversary sees at a given time for a cyber threat attack scenario depends on the intent and accumulated knowledge of the object targeted and the used cyber kill chain, which describes the stages over which a cyber threat attack scenario could transpire, to assess the opportunities, that is, the possible attack actions based on the adversary intent and accumulated knowledge. The notion of a kill chain integrated in the AABM is a set of reduction functions describing the types of attack actions that match the objective of a particular kill chain state. The type of a cyber threat attack, for instance, DDoS, data extraction, and others, or the organization describing the kill chain, the set of kill chain states may vary [28].
. In [29], the selection process of the kill chain based on fuzzy logic captures a balance between rule-based behavior models and probabilistic models. The parameters used by fuzzy logic depend on an adversary's gained knowledge of the target object and the outcomes of past actions. As described in [29], this approach allows the description of the adversary's behavior by controlling the membership functions of each kill chain as well as the definition of the kill chain outside the Minimum Viable Kill Chain (MVKC). To determine the membership for a particular kill chain a set of "attack stimuli" is generated based on input data. The set of attack stimuli used is to define the linguistic variables used in the fuzzy rules, currently represented in the AABM. This set can be separated in three categories: (1) cumulative targeted objects discovered; (2) newly discovered targeted objects; and (3) past successes and failures. These stimuli influence the definition of fuzzy rules for each of the attack stages enabling representation of an array of attacker types. The membership function is defined for each of the kill chain to describe the process by which the attacker chooses attack types. The following example shows a fuzzy rule set for an attacker type using the MVKC set.

R_1: **IF** scanned ratio is *low* **OR** newly compromised targeted object is *high* **THEN** state kill chain is *recon*

R_2: **IF** newly scanned ratio is *high* **AND** newly compromised targeted object is *low* **THEN** state kill chain is *breach*

R_3: **IF** targeted object with intent is *high* **THEN** state kill chain is *exfiltration*

For defuzzification the linguistic variables, denoted by u, they are accumulated into each of the membership functions, denoted by μ, to create a set of membership values for each of the kill chains (k_c). Thus, defuzzification is an inverse transformation, which maps the output of the aggregated fuzzy set domain back into the crisp (number oriented) output. Defuzzification can be realized by decision making algorithms that select the best crisp value based on a fuzzy set. There are several forms of defuzzification including Center of Gravity (COG), Mean of Maximum (MOM), and Center of Average Method (COAq). The COG method returns the value of the center of area under the curve and the MOM approach can be regarded as the point where balance is obtained on a curve.

The kill chain is prioritized in the order from least importance, **THEN** clause *reconnaissance*, to the most important, **THEN** clause *exfiltration*. If defuzzification is based on the maximum method it only considers active rules with the highest degree of fulfillment. Hence, the maximum of the associated output fuzzy quantity determines the sharp output size. Thus, the membership in one of the kill chains is determined by the most maximum value of the fuzzy quantities based on the basic quantity G. Then

$$\mu_1 \cup \mu_2 : G \rightarrow [0,1] \, with \left(\mu_1 \cup \mu_2 \right)(u) = MAX \left(\mu_1(u), \mu_2(u) \right)$$

which is the union of fuzzy sets μ_1 and μ_2, represented by the Maximum operator.

The reduction function from the kill chain returned from the defuzzification process is then applied to A. The set A now represents the available a with the contribution of the attacker's intent and the opportunities [29].

To demonstrate the impacts that specific types of adversaries will have on cybersecurity of a targeted object and how the configuration of the targeted object affects the success of the adversary, simulations should be executed to simulate targeted object configurations with regard to adversary types and behaviors. There are different tools available for this purpose such as the Cyber Attack Scenario and Network Defense Simulator (CASCADES), the Network Security Simulation (NeSSi), XM Cyber, Threatcare, and others.

References

1. S. Gordon, R. Ford, On the definition and classification of cybercrime. J. Comput. Virol. **2**(1), 13–20 (2006)
2. Department of Homeland Security's Role in Critical Infrastructure Protection Cybersecurity, GOA-05-434, 2005
3. https://ics-cert.us-cert.gov/content/cyber-threat-source-descriptions
4. X. Lin, P. Zavarsky, R. Ruhl, D. Lindskog, Threat modeling for CSRF attacks, in *Proceedings of the IEEE 16th International Conference on Computational Science and Engineering*, vol. 13, 2009, pp. 486–491
5. C. Phillips, L.P. Swier, A graph-based system for network-vulnerability analysis, in *Proceedings of the Workshop on New Security Paradigms*, 1998, pp. 71–79. http://doi.acm.org/10.1145/310889.310919
6. B. CSchneier, Attack trees. Doobs J. **24**(12), 21–29 (1999)
7. M. Mulazzani, S. Schrittwieser, M. Leithner, M. Huber, E.R. Weippl, Dark clouds on the horizon: using cloud storage as attack vector and online stack space, in *UNISiX Security Symposium*, 2011, pp. 65–76
8. K.P. Mandhala, J.M. Wing, An attack surface metric. IEEE Trans. Softw. Eng. **37**(3), 371–386 (2011)
9. S. Callagirone, A. Pendergast, C. Betz, The Diamond Model of Intrusion Analysis, DTIC Document, Technical Report, 2013
10. Joint Tactics, Techniques, and Procedures for Joint Intelligence Preparation of the Battlefield, U.S. Joint Chiefs of Staff, 2000

11. E.M. Hutchins, M.J. Cloppert, R.M. Amin, Intelligence-driven computer network defense informed by analysis of adversary campaigns and intrusion kill chains, in *Leading Issues in Information Warfare and Security Research*, vol. 1, (Academic Publ. International, Reading, 2011), p. 80 ff

12. H. Al-Mohannadi, Q. Mrza, A. Namanaya, I. Awan, A. Cullen, J. Disso, Cyber-attack modeling analysis techniques: an overview, in *Proceedings of the 4th International Conference on Future Internet of Things and Cloud Workshops*, 2016, pp. 69–76

13. D.J. Bodeau, C.D. McCollum, D.B. Fox, *Cyber Threat Modeling: Survey, Assessment, and Representative Framework* (Homeland Security Systems Engineering and Development Institute, Cambridge, MA, 2018)

14. OWASP Top 10 Application Security Risks-2017, 2017. https://www.owasp.org/index.php/Top_10-2017_Top_10

15. M.S. Idrees, Y. Roudier, L. Apvrille, Model the system from adversary viewpoint: threats identification modeling, in *Intrusion and Prevention Workshop*, ed. by J. Garcia-Alfana, G. Gür, In Electronic Proceedings in Theoretical Computer Science, (2014), pp. 45–57

16. A. Texeira, D. Perez, H. Sandberg, K.H. Johansson, Attack models and scenarios for networked control systems, in *Proceedings of the ACM HiCoNss*, 2012, pp. 55–63

17. H. Mouratidis, P. Giorgini, G. Manson, Using Security Attacks Scenarios to Analyze Security During Information Systems Design. http://dit.unitn.it/~pgiorgio/papers/ICEIS04.pdf

18. I. Kotenko, D. Doynikova, The CAPEC based generator of attack scenarios for network security evaluation, in *Proceedings of the IEEE 8th International Conference on Intelligent Data Acquisition and Advanced Computing Systems*, (IEEE Publ., Piscataway, NJ, 2015), pp. 436–441

19. B. Wang, J. Chai, S. Zhang, A network security assessment model-based based on attack defense game theory, in *Proceedings of the IEEE International Conference on Computer Application and System Modeling*, (IEEE Publ., Piscataway, NJ, 2010), pp. 634–639

20. Q. Do, B. Martini, K.-K.R Choo, The role of the adversary model in applied security research, in *Computers and Security*, 2018, pp. 156–181. https://eprint.iacr.org/2018/1189.pdf

21. R. McKemmish, When is digital evidence forensically sound? in *Advances in digital forensics IV*, ed. by I. Ray, S. Shenoi, (Springer Publ., New York, 2008), pp. 3–15

22. P. Liu, W. Zhang, M. Yu, Incentive-based modeling and inference of attacker intent, objectives, and strategies. ACM Trans. Inf. Syst. Security **8**(1), 78–118 (2005)

23. S. Myagmar, A.J. Lee, W. Yurcik, Threat modeling as a basis for security requirements, in *Symposium Ion Requirements Engineering for Information Security*, 2005, pp. 1–8

24. D. Grunewald, M. Lützenberger, J. Chinnow, Agent-based network security simulation, in *Proceedings of the 10th International Conference on Autonomous Agents and Multiagent Systems*, (International Foundation for Autonomous Agents and Multiagent System, Taipei, 2011), pp. 1325–1326

25. S. Moskal, B. Wheeler, D. Kreider, Context model fusion for multistage network attack simulation, in *Proceedings of the IEEE Military Communications Conference*, (IEEE Publ., Piscataway, NJ, 2014), pp. 158–163

26. U.S. Department of Homeland Security—Cybersecurity Strategy, 2018

27. J.T. Darwin, Cyberhealth and Informal Wellbeing, PhD, University of Darwin, 2019

28. S.F. Moskal, Knowledge-Based Decision Making for Simulation Cyber Attack Behaviors, PhD Thesis, Rochester Institute of Technology, 2016

29. S.F. Moskal, S.J. Yang, M.E. Kuhl, Cyber threat assessment via attack scenario simulation using an integrated adversary and network modeling approach. J. Defense Model. Simul. **15**, 13–29 (2017)

Chapter 7
Cybersecurity Ontology

7.1 Introduction

The rapid growth in data through today's digital technologies expands the importance of cybersecurity with regard to the increase of cybersecurity threats, because data are the most important value in the digital world. However, public and private organizations are currently coping with cybersecurity issues without collaboration due to lacks of global standards to solve this problem. Albeit some public and private organizations possess some forms of standards trying to solve this problem based on these standards, which cannot be deployed to fully collaborate with each other. This requires developing ontologies for cybersecurity issues which provides a common understanding of cybersecurity domains. The term ontology itself comes from the Greek words onto, which means existence or being real, and logia, which means science, or study.

Hence, the term ontology specifies some sort of shared understanding. In a more formal sense ontology can be assumed representing some kind of description logic. Furthermore, ontology may indicate that certain object types are subsets of another, and also indicate what can be said about the objects in the respective domains. As an outcome, the ontology can specify which properties each object has, and what value or range of values each property can take. In this regard ontology defines the discourse about that object. Against this background, ontology is a description of what exists specifically in a specific domain, for instance, every component that exists in an information system. This includes the relationship and hierarchy between these components. In this regard the ontology focus is not primarily discussing whether these components are the true essence or core of the information system or not. Furthermore, it is important to note that ontology does not describe whether the components within the information system are more real compared to the process that takes place within the information system. Rather, they are naming components and processes and grouping similar

D. P. F. Möller, *Cybersecurity in Digital Transformation*, SpringerBriefs on Cyber Security Systems and Networks, https://doi.org/10.1007/978-3-030-60570-4_7

ones together into categories. The purpose of ontology is to understand and describe underlying structures that affect the domain specific components or systems. In this context ontology of a domain specifies the domain specific object, concepts and relations in that domain, which can be assumed as a generally structured description of items. Hence, ontology may also indicate that certain object types are subtypes of another, and specify, which properties each object has, and what value or range of values each property can take. Therefore, ontology of a domain defines the discourse about the domain, and if an item does not appear in ontology, then about that item no statement can be given. In this context, ontology specifies some sort of shared understanding of a domain. In other words the term ontology can be assumed analogous to description logic. Some of the major characteristics of ontologies are that they ensure a common understanding of information and that they make explicit domain assumptions. As a result, the interconnectedness and interoperability of the model make it invaluable for addressing the challenges of accessing and querying data.

7.2 Ontology Types

Ontology is a formal, explicit specification of a shared conceptualization in which the knowledge model can be built upon the following types:

- *Entity*: Represents an object or thing, for example: person, smartphone manufacturer, smartphone user, and many others.
- *Relation*: Represents the relationships between entities, for example: a smartphone manufacturer and smartphone user customer relationship.
- *Role*: Describes the participation of entities in a relation, for example: in a business deal there are roles of manufacturer and user, respectively.
- *Resource*: Represents the properties associated with an *entity* or a *relation*, for example: a name or date, and others. Resources consist of primitive types and values, such as strings or integers.

Against this background ontology specifies the objects, concepts, and relations within the respective domain, and hence can be stated as a structured list of items. In this regard it is a formal naming and definition of types, properties, and interrelationships of the entities that really or fundamentally exist for a particular domain of discourse. Moreover, ontology may indicate that certain object types are subtypes of another. Hence, ontology of a domain defines the discourse about that domain. However, if an item does not appear in ontology, then that item cannot be reasoned about. In this regard ontology of a domain specifies one important type of knowledge, for instance, knowledge of the static data in the domain. This includes a vocabulary of terms, definitions of these terms, and a specification of the terms and concepts interrelations. To this extent, ontology specifies some sort of shared understanding of a domain [1]. Hence, ontologies are defined for particular purposes and in particular contents, and the form ontology takes will be at least partially influenced by those purposes and contexts [2]. Moreover, understanding of appropriate domain ontology is a great aid to knowledge acquisition. Thus, ontologies have been designed with different levels of specificity [3].

In recent years, there has been a need to use ontologies in cybersecurity for helping to solve the cybersecurity problem. In [4], the use of Semantic Web Languages and Ontologies (SWLO) for cybersecurity awareness is discussed. Hence, ontologies for cybersecurity go back to the early days of Semantic Web. For instance, in [5] the use of DARPA Agent Markup Language (DAML), the precursor of the Web Ontology Language (OWL) for representing ontology for intrusion detection issues, is discussed. It compares DAML against XML and discusses the inadequacy of the latter. The ontology includes 23 classes and 190 properties/attributes. OWL is a semantic web computational logic-based language, designed to represent rich and complex knowledge about things and the relations between them. It also provides detailed, consistent and meaningful distinctions between classes, properties and relationships. By specifying both, object classes and relationship properties as well as their hierarchical order, OWL enriches ontology modeling in semantic graph databases, also known as Resource Description Framework (RDF). RDF is a model for data publishing and interchange on the Web standardized by the World Wide Web Consortium (W3C). In this regard, RDF triplestore is a type of graph database that stores semantic facts. OWL, used together with the OWL reasoner in RDF triplestores, enables consistency checks to find any logical inconsistencies, and ensures satisfiability checks to find whether there are classes that cannot have instances. The data in a RDF triplestore is stored in three linked data pieces which are called a triple. Triples are also referred to a statement or RDF statements [6]. Also, OWL is equipped with means for defining equivalence and difference between instances, classes and properties. These relationships help users match concepts even if various data sources describe these concepts somewhat differently. They also ensure the disambiguation between different instances that share the same names or descriptions [6].

7.3 Cybersecurity Ontology

The rapid growth in data through today's digital technologies expands the importance of cybersecurity with regard to the increase of cybersecurity threats, because data is the most important value in the digital world. In this context, this data is available in structured, semi-structured, and unstructured forms for both, data from internal and external sources. Therefore, unifying such scattered data will provide better visibility and situational awareness with regard to cybersecurity analysis as well as a more proactive and possibly predictive approach to avoid cyber threats. Against this background the development of cybersecurity attack ontology is essential to enable the secure data integration across disparate data sources. In this context cybersecurity attack ontology is required for modeling different types of adversary knowledge. Therefore, security attack ontology aims at building a knowledge base for security attacks that describe type, mode, consequences, and others. Developing the cyberse-curity attack ontology one can make use of known security standards (see Sect. 6.3), for instance ISO/IEC 15408:2009, ISO/IEC 18045, ISO/IEC 27000:2012, ISO/IEC 17799:2005, NIST SP-800:30, and others, and security dictionaries (see Sect. 6.3), for instance Common Vulnerabilities and Exposures (CVE), CAPEC™, OWASP, Comprehensive Lightweight Application Security Process (CLASP), and others. Thus, the security ontology can make use of the foregoing constructs with regard to the Web Ontology Language (OWL), a language for defining ontologies to describe properties of web resources [7]. OWL is a semantic web computational logic-based language, designed to represent rich and complex knowledge about things and the relations between them. It also provides detailed, consistent and meaningful distinc-tions between classes, properties and relationships. OWL based ontology describes a domain in terms of classes, instances and relations and include descriptions of the characteristics of those objects with regard to slots and internal links such as instance-of and subclass-of. Based on the conceptual aspects about attack models and attack scenarios presented in Chap. 6 and the security standards, the cyber security attack ontology can be illustrated as shown in Fig. 7.1.

In the context of semantics it is possible to execute precise searches and complex queries. Initially, this effort is focused on cyber threat malware subjects, because malware is one of the most prevalent cyber threats to cybersecurity. For this reason the MITRE Corporation has developed the Malware Attribute Enumeration and Characterization (MAEC) language [8] which is a structured language for encoding and sharing high-fidelity information about malware based upon attributes such as behaviors, artifacts, and relationships between malware subjects. As described in [8], MAEC focuses on characterizing the most common malware types, including Trojans, worms, rootkits, and many others, as well as today's more advanced mal-ware types. MAEC's core components include a vocabulary, a grammar, and a stan-dardized output format, and provide a standard means of communicating information about malware attributes, as shown in Fig. 7.2.

Before MAEC, the lack of an accepted standard for unambiguously characteriz-ing malware subjects meant there was no clear method for communicating the spe-cific malware attributes detected in malware by the analyses, or for enumerating its fundamental makeup. The results included non-interoperable and disparate malware reporting between public and private organizations, disjointed or inaccurate mal-ware attribution, the duplication of malware analysis efforts, increased difficulty in determining the severity of a malware threat, and a greater delay between malware infection and detection as well as response [9]. However, the key to ontology devel-opment is an understanding of the cyber domain, which drives the kinds of entities, properties, relationships, and potentially rules that will be needed in the ontology. With regard to the complexity of cybersecurity analysis, the ontology development better consist of modular sub-ontologies, rather than a single, monolithic ontology [10]. Thus ontologies can be grouped categories such as upper level ontologies, mid-level ontologies, and domain level ontologies, according to their specific levels of abstraction of the respective cybersecurity architecture ontology concept to be developed. For more details see [10–12].

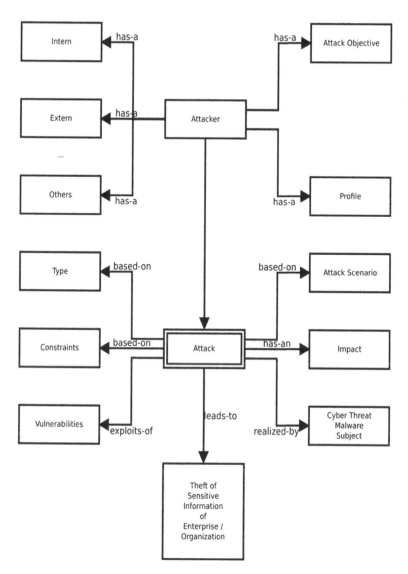

Fig. 7.1 Cyber security attack ontology

Developing the detailed architecture of the cybersecurity ontology requires, dependent of the category of interest, specific descriptions to abstract major categories, domain specific concepts, and ontologies that span multiple concept categories. The descriptions of the major categories which lay the basis for a cybersecurity ontology taxonomy are:

- *Entities*: Describe foundational incidents, collections, and others.

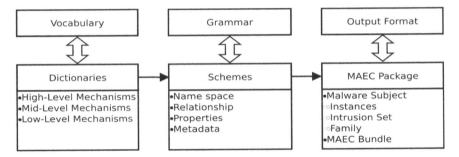

Fig. 7.2 MAEC's core components vocabulary, grammar, and output format

- *Relations*: Describe relationships of detection and defense actions, organizational locations, and others.
- *Role*: Describe cyber threat attackers and cyber threat defenders.
- *Resources*: Describe capability, infrastructure, behavior, malware subjects, and others.

In regard to malware, resources published that attempt to systematically categorize malware subject's ontology are reported in [13], and descriptive languages implemented in Extensible Markup Language (XML) in [7, 14, 15]. The ontology described enables data exchange between security algorithms. Their taxonomy of malware classes is shown in Fig. 7.3. Also, worthy to mention is an attempt of categorizing malware subject traits [16]. This development finally ended up in the so-called Unified Cybersecurity Ontology (UCO) framework described in [16] helping to evolve cybersecurity standards from a syntactic representation to a more semantic representation showing several contributions for the cybersecurity ontology. In this regard UCO is an extension to Intrusion Detection System ontology [5] to describe incidents related to cybersecurity. Several projects that focus on individual components of a Unified Cybersecurity Ontology framework analyze different data streams and assert facts in a so called triplestore approach, as reported in [5, 17, 18]. In this context UCO is essential for unifying information from heterogeneous sources and supporting reasoning and rule writing. Thus, UCO supports reasoning and inferring new information from existing information, and also supports capturing specialized knowledge of cybersecurity analysts which can be expressed using ontology classes and terms as well as rules.

Besides OWL language, the MITRE Corporation has launched the Malware Attribute Enumeration and Characterization (MAEC) language [7], a structured language for encoding and sharing high-fidelity information about malware subjects based upon attributes such as behaviors, artifacts, and relationships between malware subjects. Malware is responsible for a variety of malicious activities, ranging from spam email distribution via botnets to the theft of sensitive information via targeted cyber threat attacks. Therefore, the protection of computer systems and networks from malware is a primary cybersecurity concern for public and private organizations, as even a single instance of uncaught malware can result in damaged

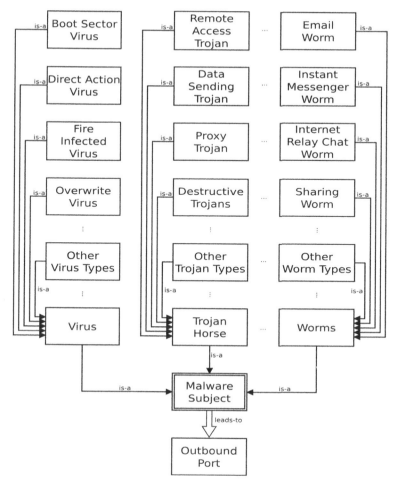

Fig. 7.3 Taxonomy of malware classes

computer systems and compromised data. However, the key to ontology development is an understanding of the respective cyber domain, which drives the kinds of entities, properties, relationships, and potential rules essential to the cybersecurity ontology.

Against this background the cybersecurity ontology framework includes the cybersecurity domain-specific ontology and data integration for different data sources in a common knowledge base, for instance, metadata records. This enables data integration and padding from ontology information and access to various data sets. This also has to include security services related to the respective business processes, network devices, and the requirements ultimately required to provide cybersecurity against cyber threat attack incidents as part of the middleware. The integration is the required interaction between data set infrastructure and cybersecurity ontology layers that provide the requirements for cybersecurity to detect cyber threat attack incidents, prevent cyber threat attack incidents, and avert cyber threat attack incidents for developing cybersecurity domain-specific ontology.

With respect to cyber threat attack intrusion incidents on data sets, it is assumed that a cyberattack maps to the category unknown, pointing to unpredictable and unexpected cyber threat attack incidents. This represents a dynamically changing risk for the data space, in the digital transformation era, which requires an adequate solution for unpredictable incidents to make data space cyber-secure. This requires the domain-specific semantics of unknowns as a kind of uncertainty that must be represented by their ontologies. Such ontologies must be able to suggest suitable cybersecurity services that may or may not be required, which have to be set at design time of the data record, and customized and activated by the data sets used. Hence, the architecture of a generic cybersecurity ontology framework is based on components, as shown in the generic model in Fig. 7.4.

The generic cybersecurity ontology framework in Fig. 7.4 shows the essential system components, including the cybersecurity domain-specific ontology and data integration for different data sources in a common knowledge base, for instance

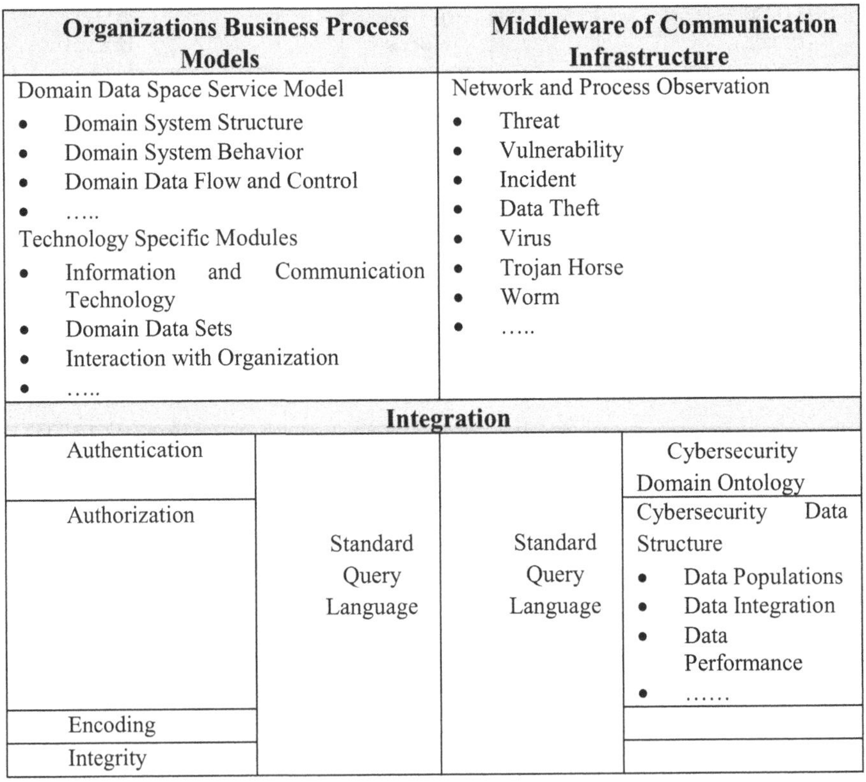

Organizations Business Process Models			Middleware of Communication Infrastructure	
Domain Data Space Service Model			Network and Process Observation	
• Domain System Structure			• Threat	
• Domain System Behavior			• Vulnerability	
• Domain Data Flow and Control			• Incident	
•			• Data Theft	
Technology Specific Modules			• Virus	
• Information and Communication Technology			• Trojan Horse	
• Domain Data Sets			• Worm	
• Interaction with Organization			•	
•				
Integration				
Authentication			Cybersecurity Domain Ontology	
Authorization	Standard Query Language	Standard Query Language	Cybersecurity Data Structure	
			• Data Populations	
			• Data Integration	
			• Data Performance	
			•	
Encoding				
Integrity				

Fig. 7.4 Generic cybersecurity material data space ontology framework

metadata sets. This enables data integration and padding from ontology information and access to various data sets required. Furthermore, security services related to organizations business process models, network devices, and the requirements ultimately required providing security against cyber threat incidents.

Data sets integration in the middleware layer provides the requirements for cyber security, for instance cyber threat attack intrusion incidents and other vulnerabilities that create the security framework for using domain-specific cybersecurity ontologies.

The queries combining all values of data sets that are analyzed for cyber threat attacks using cybersecurity domain-specific context ontologies. The queries for identifying possible cyber threat attack incidents according to the architecture shown in Fig. 7.4 must, as described, characteristically map in depth the underlying information of the considered data sets in ontologies so that cybersecurity can map domain-specific ontologies to it.

The cybersecurity core ontologies form, in a certain sense, strengths and weaknesses profiles that map the security requirements to the possible entities. The ontology for cyber secure operations aims to reduce potential false positives in detecting potential cyberattacks that may arise when monitoring cyber vulnerabilities. Thus, cybersecurity ontology represents a domain-specific model that defines the essential domain concepts, their properties and the relationships between them and represents an essential knowledge base to cyber secure the respective application. The generic cyber-attack model is shown in Fig. 7.5.

As shown in Fig. 7.5 cyber threat analysis is a security field that needs a more scientific basis for sharing information among cyber defending teams. One option is building OWL-based malware analysis ontology to provide that more scientific approach based on a malware analysis dictionary and taxonomy, and combining those in a competency model with the goal of creating an ontology-based cybersecurity framework. Meanwhile several security standards have been developed, taking into account OWL, representing ontology based security concepts such as: incident reporting, threat information, risk information, assets, target information. Each group contains multiple metrics, also known as *factors,* used to compute a Common Weakness Security System (CWSS™) score for weaknesses. CWSS™ is co-sponsored by the MITRE Corp. [18]. Thus, ontology can be defined as abstract representation of real-world objects, which means that ontology constitutes a domain-specific model defining the essential domain concepts, their properties, and the relationships between them, represented as a knowledge base.

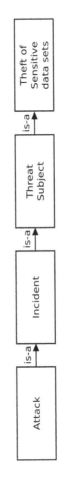

Fig. 7.5 Cyber-attack model

References

1. M. Uschold, Knowledge level modeling: concepts and terminology. Knowl. Eng. Rev. **13**, 5–29 (1998)
2. B. Chandrasekaran, J.R. Josephson, V.R. Benjamins, The ontology of tasks and methods, in *Proceedings of the 11th Banff Knowledge Acquisition for Knowledge for Knowledge-Based System Workshop*, 1998
3. N. Sadbolt, K.O. Hara, H. Cottam, The use of ontologies for knowledge acquisition, in *Knowledge Engineering and Agent Technology*, ed. by J. Cuena, Y. Demazeau, A.G. Serrano, J. Treur, (IOS Press, Amsterdam, 2004), pp. 19–42
4. A. Sheth, Can semantic web techniques empower comprehension and projection in cyber situational awareness, in *ARO Workshop*, 2007
5. J. Undercoffer, J. Pinkston, A. Joshi, T. Finn, A target-centric ontology for intrusion detection, in *18th International Joint Conference on AI*, 2004, pp. 9–15
6. https://www.ontotext.com/knowledgehub/fundamentals/what-are-ontologies/
7. S. Bechhofer, OWL: web ontology language, in *Encyclopedia of Database Systems*, ed. by L. Liu, M.T. Özsu, (Springer Publ., New York, 2009). https://doi.org/10.1007/978-0-387-39940-9_1073
8. MAEC—Malware Attribute Enumeration and Characterization. http://maec.mitre.org/
9. http://maecproject.github.io/about-maec/
10. L. Obrst, P. Chase, R. Markeloff, Developing an Ontology of the Cyber Security Domain. http://ceur-ws.org/Vol-966/STIDS2012_T06_ObrstEtAl_CyberOntology.pdf
11. L. Obrst, Ontolological architectures in Theory and Applications of Ontology - Computer Applications, ed. by J. Seibt, A. Kameas, R. Poli, Chapter 2, pp. 27–66, (Springer Publ. London, 2010)
12. S. Semy, M. Pulvermacher, L. Obrst, Toward the Use of an Upper Ontology for U.S. Government and U.S. Military Domains: An Evaluation, MITRE Technical Report, MTR 04B0000063, 2005
13. M. Swimmer, Towards an Ontology of Malware Classes. http://www.scribd.com/doc/24058261/Towards-an-Ontology-of-Malware-Classes
14. IEEE-SA—Industry Connections. http://standards.ieee.org/develop/indconn/icsg/malware.html
15. MANDIANT: Intelligent Information Security. http://www.mandiant.com
16. L. Zeltser, Categories of common malware traits, Internet Storm Center Handler's Diary, 2009. http://isc.sans.edu/diary.html?storyid=7186
17. S. More, M. Matthews, A. Joshi, T. Finn, A knowledge-based approach to intrusion detection modeling, in *Proceedings of the IEEE Symposium on Security and Privacy Workshops*, 2012, pp. 75–81
18. https://cwe.mitre.org/cwss/cwss_v1.0.1.htm

Chapter 8
Cybersecurity Leadership

8.1 Introduction

Digitalization is based on advanced digital technologies and the accompanying digital transformation of economic and social structures and processes which represent a central challenge for corporate development. Therefore, fundamental changes of economic structures and processes require a fundamental expansion of the management perspective in the era of digital transformation in the direction of digital corporate management. Hence, for their main fields of action, a far-reaching and continuous increase in the degree of digitization can be assumed for the future [1]. However, apart from digital transformation a new form of cyber threat attacks and vulnerability come into action which can attack digital technologies and hence digital corporate management. The result is reputational harm and destruction of data, and increase in cost of defending public and private organizations data in the virtual data space. The way of navigating this global virtual data space, directly affects the real life, economy, politics, businesses, and can potentially have dire consequences. Therefore, public and private organizations have to follow the digital and technological advancements, to be prepared for the digital transformation and gain critical knowledge on emerging trends in cybersecurity to become a leader in cybersecurity.

8.2 Roadmap to Cybersecurity Leadership

Cybersecurity methods to defend cyber threat attacks on systems and networks require an intelligent risk management to balance limited available resources against the need to secure organizations from ever evolving cyber threat attacks incidents in the age of digital transformation. However, cybersecurity problems are multifaceted, which cannot be solved with a one dimensional approach. Hence, multi-dimensional

D. P. F. Möller, *Cybersecurity in Digital Transformation*, SpringerBriefs on
Cyber Security Systems and Networks, https://doi.org/10.1007/978-3-030-60570-4_8

and multidisciplinary knowledge is required to become a cybersecurity leader. The reason for that is, that a cybersecurity leader has to understand and dominate the strategy of an organization's business and its vulnerabilities with regard to cyber threat attacks. Risk management and strategic prioritization of expenditures for cybersecurity tools as well as cybersecurity defense teams are key features permeating such a strategy. Hence, cybersecurity leaders focus on opportunities how to defend cyber threats. Thus, a balanced implementation of cybersecurity increases organizations productivity and innovation. Hence, cybersecurity leadership is a business discipline, as stated in [2], because senior executives also need to view technology as a core component of their business.

Today's economy and society is driven by digital transformation processes creating a new world based on software. Software will come with software updates with new capabilities to do new things, a transformation of basic features or things. This raises a real challenge in digital technologies and hence requires capabilities in cybersecurity issues as well as digital assets turning digital technologies into an organizational transformation. This requires leadership competencies that are leadership skill and behaviors that contain [2].

- Decision making competencies
- Diversification of competencies merging young and elderly employees
- Ethical competences
- Organizational Competencies
- Technological competencies

To name a few. This makes risk management better, faster, and will result in more successful decisions, and therefore, security leadership to reduce wasting time presuming alerts based on:

- Cyber threat attack incidents that are more likely to be innocuous rather than malicious
- Cyber threat attack incidents that are not relevant to the organization
- Cyber threat attack incidents for which defense algorithms and controls are already in place.

Against this background, Cybersecurity Leadership (CSL) has to have the competencies mentioned above to:

- Assess technical organizational risks, including emerging threats and "known-unknowns" that might impact an organization (see Table 4.6)
- Identify the right strategies in cyber threat attack security to mitigate the risks
- Communicate the risk nature to the top management (CIO) of the organization to justify investments in defensive measures

However, threat intelligence methods can become a critical resource for all these action items, providing information on general trends, such as:

- Which types of cyber threat attacks are becoming more (or less) frequent

- Which types of cyber threat attacks are the most costly for the attacked public and private organization
- What new kinds of cyber-attackers are coming forward, and what are the assets and public and private organizations they are targeting
- What are the security practices and technologies that have been proven the most (or least) successful ones in stopping or mitigating the respective cyber threat attacks
- And much more

With this knowledge, gathered from a broad set of external data sources, security decision leaders being able to gain a holistic view of the cyber risk landscape and the greatest risk potential that may happen to their organization. In this regard, cybersecurity leadership can be achieved referring to the main core areas where threat intelligence (see Chap. 3) helping cybersecurity leaders making successful decisions to perform a well trusted cybersecurity leadership.

8.3 Digital Master

Digitization is a top priority for organizations through the digital transformation with its torrent of data generated and its impact on the transformation process. However, the dynamics of the current developments in digital transformation make digital business development a constant task. Hence, digital masters will take this huge amount of data, combine it with the latest innovations in Artificial Intelligence, Machine Learning, Internet of Things, Big Data and Analytics, and others, and use the resulting insights to make smarter decisions, see the future more clearly, and drive out inefficiencies with regard to digital businesses and organizational development. Therefore, a digital master has to understand the digital upheavals and innovations to develop future-proof digital business strategies, design structures, marketing concepts and data and process security concepts for public and private organizations. Thus, digital masters have to combine digital capabilities and leadership capabilities to achieve performance that is greater than either dimension can deliver on its own. Hence, digital capabilities make new digital initiatives easier and less risky, for the digital masters and providing revenue leverage to transform digital business [1]. Albeit digital capabilities support the focus on the uniqueness of organizations digital master's strategy with a view to a cross-company digital business development should add value and competitiveness to this uniqueness. Hence, digital masters rely on data-driven insights to decide how to proceed to drive efficiency. This requires building a governance model with its respective committees that suits organization needs. Based on that, a governance model can be built that fits with essential needs of the organization. However, the steps before require deciding how to develop digital leadership/mastering capabilities of the public and private organization, the key to innovation and sustained value creation. This is based on an innovative blueprint for the respective public and private organizations' business models to embrace the wave of digital transformation.

References

1. J. Schellinger, K.O. Tokarski, I. Kissling-Näf, From digital transformation to digital corporate management, in *Digital Transformation and Corporate Management*, ed. by J. Schellinger, K.O. Tokarski, I. Kissling-Näf (Springer-Gabler, Wiesbaden, 2020), pp. 1–10 (in German)
2. M. Hasib, *Cybersecurity Leadership—Powering the Modern Organization* (Tomorrows Strategy Today, LLC Publ., 2015, U.S.A.) ISBN 978-1496199270

Printed in the United States
By Bookmasters